Management
of
Engineering Design

Too big, too expensive, too late

AWRE Scientist

The optimum solution to the sum of the true needs of a particular set of circumstances

Matchett

An ill-assorted collection of poorly-matching parts forming a distressing whole

Datamation

Definitions of Engineering Design supplied by G. J. Terry of the A.W.R.E.

Management
of
Engineering Design

D. J. LEECH

Industrial and Systems Engineering Division,
University College of Swansea

JOHN WILEY & SONS

London – New York – Sydney – Toronto

Library of Congress Catalog Card No. 73-37112

ISBN 0 471 52270 8

Printed in Belgium by
Ceuterick, printers since 1804
B 3000 Louvain

PREFACE

This book has developed from a course of 24 lectures given to 2nd year students in a University Engineering School. The original motivation for the lectures was provided by the author's experience in running technical departments in industry which revealed that many engineers have little appreciation of strategy in design.

Part I discusses the need for formal problem definition and develops a design specification. Most firms have found the need for a formal approach to the writing of design specifications although, in practice, many designers still resist the discipline of formal problem statements. In industry the design specification is an essential document which must be recognized by both customer and designer but, in college, the design specification is also a useful peg on which to hang discussions of objectivity, economics and environment. The design specification evolved in Chapter 7 is rather formal. This is for two reasons; firstly, the author's experience has been in a field (aircraft accessories) in which design specifications are usually the bases for contractual agreements but secondly because undergraduate students seem to prefer, and learn more from, the formal approach. It should be pointed out that some workers* do not like very formal specifications because they find that they limit creativity.

Part II discusses aids to creativity and should enable students to hold 'brainstorming' or 'synectic' sessions.

Much engineering is taught as though it were applied mathematics and this makes the student suspicious of any approach that is not formally mathematical. It is clear that much of the work on strategy cannot be entirely quantitative and while there is a perfectly respectable science of decision making, in real life this science cannot often be applied. An engineer learns by experience, which is another way of saying that he learns by his mistakes and, where formal rules cannot be taught, it is often possible to teach by case histories of the mistakes of others. Examples are used to make points throughout the book but the author has found it useful to invite practising engineers to discuss with students the histories of real designs that failed. The other method of teaching strategy in design is by project work and it cannot be overemphasized that a student learns much more by trying to solve a real problem than by hearing lectures. The exercises on Parts I and II may form

* E.g. E.M. Matchett (Engineering Employers' West of England Association).

the basis of extensive project work but it is always more informative if the objective approach can be developed with real problems obtained from industry.

Part III was written because in industry many designers seem not to realize that what they draw has to be made, and made to work, by other men in the factory. Since many practising engineers seem not to understand the context in which they work it is unlikely that an undergraduate will have a prior knowledge of the structure of the team of which he will one day be a member. Some attempt has been made to formalize this Part by defining the paperwork which has to be used in a typical design organization.

Part IV has been written because it is a good way of making the designer aware of the importance of time and money. It would not normally have been thought necessary to write at length on Critical Path Methods, because there are plenty of texts at the elementary level, but in real life even the most elementary network problem founders unless resources have been taken into account, and this is not done by most available texts. It was therefore thought necessary to write elementary introductions to network analysis with resource limitations and to the use of networks in cost control.

Part V consists of a number of chapters which are not necessarily related to one another. Chapter 16 has been included because there appears to be no satisfactory, small, cheap test on Reliability, to which the student can be referred and because the subject is an important one which must be considered in the course of every design. Chapters 17, 18 and 19 are intended to introduce more quantitative methods of decision making. These quantitative techniques help to drive home the strategic approach to design and also give a little insight into one of the fashionable aspects of Computer Aided Design. However, dealing with optimization in any depth would require a complete course and a whole book. It was therefore decided that only one optimization technique would be taken to a stage at which the student could attempt much calculation. Linear Programming was therefore given a chapter which leads to an explanation of the Simplex method by largely intuitive methods. Classical methods of optimization are usually dealt with in other courses while a deeper study of Hill-Climbing and Dynamic Programming is best deferred until the student has some computing experience and time for extended optimization exercises.

The discussion of Value Engineering has been included as an appendix. The reason for doing this is that the whole approach to design that has been used is that of the Value Engineer and the appendix has been included to draw this parallel, rather than teach techniques. Techniques are discussed, incidentally, but generally one would expect Value Engineering, as a job, to be something for the man with many years' experience.

While the book is intended as an undergraduate course, it is hoped that it will be of value and interest to the practising designer.

ACKNOWLEDGMENTS

Many people have been helpful in supplying illustrations for the text:
    Teddington Aircraft Controls of Yeovil, Somerset, provided almost all the illustrations for Chapter 12 as well as much help, particularly for examples and on the subject of cost control.

The Atomic Weapons Research Establishment provided information and illustrations of the PABLA system and the remarks of Mr. G. J. Terry of that establishment have been very helpful.

Professor Kirk of Loughborough University of Technology gave permission for the reproduction of Figure 20.2 and the information associated with it.

Undergraduates in the division of Industrial and Systems Engineering at Swansea have supplied much information and typical of their work are Figures 16.9 and 16.11. The matrix of the Weibull graph paper and the confidence level curves were plotted by computer using programs written as undergraduate exercises.

The Society of British Aerospace Companies provided Figure 12.3.

The extract from B.S. 2G 100, General Requirements for Electrical Equipment and Indicating Instruments for Aircraft, Part 2: Environmental and Operating Conditions, is reproduced by permission of the British Standards Institution, 2 Park Street, London W1A 2BS.

CONTENTS

PART 4 THE MANAGEMENT OF DESIGN

PART 5 MODELS AND TOOLS OF ANALYSIS

Part 1
DEFINING THE PROBLEM

If you walk through a design office you may well find a number of earnest hard-working designers who, with 4H pencils, are getting to close grips with problems that they have not defined and do not understand.

It often seems that many designers to not realize that they are working in a commercial field or that their designs must work in the real world, and they expect to be judged by their abilities at producing trivially ingenious mechanisms. The designer must consider not only the purpose of his design but also the economic implications of his work and the environment in which his design will be expected to operate. The result of this consideration is the design specification. The first seven chapters of this book will have succeeded if they convince the intending designer of the importance of (a) costs, (b) environment and (c) formal design specifications.

The points are made with the aid of examples, all of which are based on reality although, for obvious reasons, simplifications have been made and values have been altered. One example, a pressure switch, is pursued in rather more depth than the others.

In the literature available at undergraduate level there is not a great deal about design specifications. The early chapters of Gosling (2a),* the early chapters of Chestnut, 1965, and Woodson, 1966, provide useful background reading and, in particular, Woodson provides some useful subjects for project work. The best background reading is undoubtedly the material provided by the documents actually used in industry. A specification that has been produced by one of the great engine companies of the world will teach a student far more than most text books. Part of B.S. 2G. 100†, (6b) is referred to in the text and I find it a useful exercise for the student to read the whole of part 2 of this publication. The Feildon Report, (6a), also puts design into its proper context and is essential reading for any designer.

Undoubtedly, the engineer learns from doing, more than from reading, and any course of design must include time given to the solution of problems. It is best if those problems are real. Problems, particularly those involving specification writing, require the student to ask many questions; artificial problems raise questions to which artificial answers have to be created and it is not easy for a teacher to predict likely questions and create all the necessary answers in advance. If local engineering firms can be talked into providing small projects, these will be found to have much greater value than artificially created problems. Such projects will also be much more likely to create that excitement in the student that is so much part of the good designer.

* (2a) implies that reference is made to this text in Chapter 2. The reference is listed in the bibliography. Other references are listed in the bibliography as further reading.
† Currently being replaced by B.S. 3G. 100.

CHAPTER 1
The Design Situation

The scientist studies natural phenomena because he wishes to understand them and believes himself to be successful if his understanding is increased. The engineer also studies natural phenomena but he does this because he wishes to use the knowledge he acquires to make himself (or somebody else) richer. The engineer believes himself to be successful when his efforts result in something saleable. It is important that the engineer should realize this and should not regard himself as a second class scientist; his motives are quite different.

The engineering giants of the past were not frustrated scientists. Watt improved the steam engine and formed a company to exploit his patents. Brunel was employed to build a railway. Whittle obtained support to design and build a gas turbine to power R.A.F. aircraft. These men, no doubt, learnt a great deal that was new about thermodynamics, structures, metallurgy etc. but their customers (or sponsors) wanted a steam engine, a railway or a gas turbine and would not have paid for anything less.

The engineer makes things and one thing he must make is money. This fact makes the engineer's job both difficult and exciting. In the long run an engineer cannot short-change his customer and keep in business. The customer sets the engineer an objective and it will be obvious to everybody if the engineer does not meet his obligations. If an engineer contracts to build a bridge, it will be apparent, even to the layman, if the bridge is not built, or if the bridge is built late, or if the bridge is built and then falls down.

We have spoken of the engineer as if he is one man but even the simplest analysis will consider the engineers who manufacture and the engineers who design. We will, then, consider the simple situation in which we have:

a customer who wants something and is prepared to pay for it,

a manufacturing organization which will make what the customer wants and sell it to him

and

a design engineer who supplies the manufacturer with what instructions are necessary to manufacture his product.

Although the model is very simple, most real life situations may be approximated to it.

Consider the customer. He has decided that he needs a product (or he has been convinced by the engineer that he needs it). Because he wants the product, he will be prepared to pay for it and if he is an intelligent customer he will have worked out how much the product is worth to him. The price that the customer is prepared to pay is the selling price of the product. If the manufacturer is unable to make the product at this selling price then he had better not make it at all. It does not matter how elegant the product, if the manufacturer cannot sell it at a profit, then it is not worth making. In most sophisticated engineering fields, the customer will require the product so that he in turn can make a profit from it. B.O.A.C. or Pan American will buy an aeroplane in order to make a profit out of operating it; the Central Electricity Generating Board will buy a power station in order to be able to sell electricity at a profit. Such customers will therefore calculate, very carefully, how much a new product is worth to them and will think that a good product is one that by its use makes them a satisfactory profit.

The manufacturer will think that a good product is one which he can sell at a profit. Clearly there are two different criteria by which a product is to be judged although they are not necessarily incompatible. If a manufacturer is to stay in business he must have a succession of satisfied customers. If the manufacturer is to make a product which he can sell at a profit then he must design one which will make a profit for the customer.

The designer's function in all this is to decide what the customer requires, decide what product will meet the customer's need and provide all the instructions which will enable the manufacturer to make the product. Usually the designer is employed by the manufacturer, so that a good designer will be one who makes most profit for the manufacturer—at least in the eyes of his employer. Since his employer's opinion determines the designer's salary, then this is a very important criterion to the designer.

The design situation is therefore entirely a commercial one which we may summarize by saying that:

an engineered product must meet a customer's need,

the selling price of a product is that which the customer is prepared to pay,

for the customer, the best product is that which brings him most profit,

for the manufacturer, the best product is that which brings him most profit, the designer decides what the customer needs and provides the manufacturer with the instructions necessary to make it

and the best designer makes most profit for his employer.

These are, of course, generalizations which do not apply in all cases as baldly as they are stated. We have chosen to discuss profit in very simple terms, as though profit were no more than the amount of money we take home after a week's work or the cheque we put in the bank after a successful sale. Clearly, it is much easier to derive our argument if we confine ourselves to such simple ideas, but, in reality, profit is more complex. When a man buys a car he may not expect it to bring him a profit, in the simple sense of the word, but he expects it to bring him pleasure or convenience which can be equated to sums of money. He will be prepared to make judgments between alternative qualities; he may for example, be prepared to trade some comfort for improved performance or to trade acceleration for load-carrying ability. Such judgements involve the comparison of qualities which are clearly unlike each other. To make them comparable we must assign values (utilities) to the qualities we are comparing and money is a sensible yardstick to use for the comparison of values. Profit then, although expressed in money terms, may contain allowances for many other properties thought desirable.

We cannot calculate profit without calculating cost, and here again we have to compare dissimilar properties. Part of the cost of the Concorde project, for example, is the possibility of damage to St. David's Cathedral resulting from sonic bangs. Whether consciously or not, we must attempt to assign a money penalty to such a possibility before we can decide whether to go ahead with the Concorde project or not. Certainly, among those who wish to stop the Concorde are many who believe that there is a high expectation of heavy damage by sonic bangs along the Concorde route and who are prepared to argue that this expectation of damage is equivalent to a very high cost. There are in fact, few situations in which we cannot determine the profit and loss in terms of money. It is sometimes difficult however, to analyse situations in this way because we are not always prepared, or able, to put a money value on our satisfaction or dismay and sometimes our knowledge is so imprecise that we cannot predict the probabilities associated with possible outcomes.

It is sometimes argued that saleability is not necessarily a major criterion in all cases and medical, military and space applications have been cited as situations in which the ordinary commercial customer/seller relationships do not apply. But if we consider such products as a kidney machine, a nuclear submarine or a flight to the moon we soon observe that commercial consider-ations must, and do, apply. Someone must pay for a kidney machine and recent public debate has clearly demonstrated the care with which public bodies compare the relative utilities of alternative claims on their limited financial resources. Similar debates have been observed about the claims on the public purse of different defence systems and on the United States' space programme's share of the resources available. It is true that the profit to the customer from these systems may not show directly as money but there is little doubt that, for example, a kidney machine will be considered by its user to have some value.

CHAPTER 2
The System

When we ask ourselves what products concern the engineer we realize that he is involved in almost everything that is manufactured. A product is really defined when a customer's need is established and to meet this need we may require a product of little or great complexity. If the customer wants something to put his dinner on, then he may be satisfied with a plate and while the manufacture of a plate may be a complex operation, the item itself is simple and consists of one part only. If, however, the customer wants a system which will enable people to travel from one place to another, then he may require a complete airline system to be produced. Nowadays we tend to divide engineering products into components and systems. A component may, by itself, fulfil a customer's need and be sufficiently simple for one man, with one discipline at his disposal, to design. Some customer's needs however, cannot be met by a simple component and in such cases it is necessary to design a system. A system consists of many components and its design may require the services of many men with many disciplines. As we shall see, it is necessary to regard almost every engineered product as a system at some stage in its design or analysis and a system may best be defined after Gosling, (2a), in the following way: 'A System is a product which is most easily analysed, described or designed as an assembly of smaller parts.' 'System' is a relative term since the level to which we divide a system into its component parts is arbitrarily decided by convenience of study.

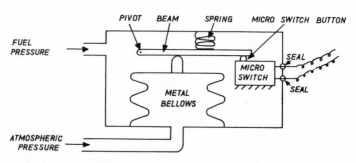

Fig. 2.1(i) A diagrammatic representation of a pressure switch

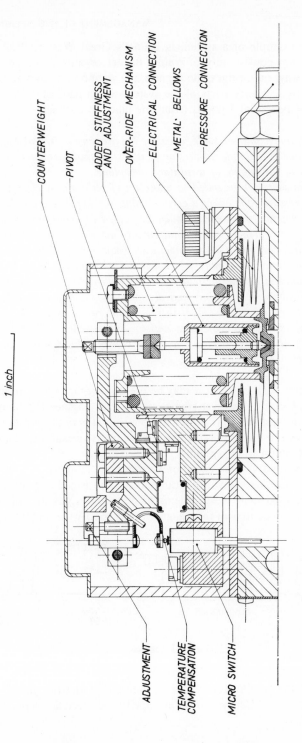

COUNTERWEIGHT

PIVOT

ADDED STIFFNESS
AND ADJUSTMENT

OVER-RIDE MECHANISM

ELECTRICAL CONNECTION

METAL BELLOWS

PRESSURE CONNECTION

1 inch

ADJUSTMENT

TEMPERATURE
COMPENSATION

MICRO SWITCH

Fig. 2.1(ii) – – – – and what the diagram becomes

An early example of a system is Brunel's Great Western Railway. This system did not consist only of trains. Brunel organized the design of the permanent way, the bridges, the tunnels, the stations etc., and all these components were necessary parts of the system which had, as its objective, the transport of people and goods from one place to another.

On a very much smaller scale, consider the pressure switch of Fig. 2.1. This pressure switch is a simplified version of a system that was supplied to meet a requirement in the engine of a military aeroplane.

The whole switch is intended to make an electric alarm circuit if the fuel (paraffin) pressure falls below a certain value but, because the switch is working in fuel vapour, the contacts must be sealed to avoid danger of a spark igniting the fuel. Further, it is a requirement that the whole pressure switch be very small and works in a high-temperature environment. The design of a very small switch contact assembly which will work at high temperatures and in which the contacts are hermetically sealed, requires specialist knowledge that the average mechanical engineering designer does not possess. The designer would therefore find it convenient to regard the pressure switch as a system consisting of the mechanical parts and the micro-switch. He will define the requirements of the micro-switch and this can then be designed by a second specialist designer.

This example suggests that it may be useful to regard quite simple designs as systems. Usually, however, the term 'systems engineering' is applied to large, complex systems—say power stations, oil refineries or weapons systems although it will be useful to return to our switch as an example of some of the problems common to all systems.

The following are some common examples of systems:

1. The gas turbine, shown diagrammatically in Fig. 2.2. A chief designer may be presented with the problems of designing a gas turbine and while he retains overall control of the design, he may specify the requirements of the

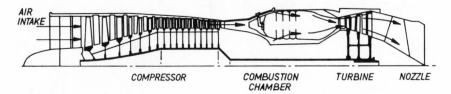

COMPRESSOR   COMBUSTION   TURBINE   NOZZLE
                    CHAMBER

Fig. 2.2 Gas turbine

compressor, the combustion chamber, the turbine and the nozzle and have each of these parts designed and developed by a specialist department. The system designer must never lose sight of the customer's objective, which in this case is probably the supply of a prime mover for his aeroplane, and eventually the whole gas turbine will be supplied to meet this objective. The component parts may, however, be largely regarded as self-contained design problems provided that the work is properly co-ordinated (in the gas turbine, only 4 sub-systems have been listed but in fact there are many more : the fuel system, the lubrication system, the auxiliary power systems etc., some of which may be designed and built by firms other than that responsible for the parent system).

2. A large weapons system may have as its centre a missile carrying an atomic war head, but peripheral to this we must have a launching site, a guidance system and many other sub-systems. Each of the sub-systems will be the reponsibility of a specialist organization but a systems engineer will be required to co-ordinate the work of all the different teams.

3. A transport system consists of many parts and we have already mentioned the railway. While the design of rolling stock, of permanent way, of stations etc., may all be jobs for specialist organizations, a systems engineer must co-ordinate the work of these separate teams.

It is clear from these examples that almost anything can be regarded as a system and the system designer's approach may be useful in solving any but the most trivial engineering problems.

Systems may be divided into two classes, according to their complexity. Some of the simpler systems may be analysed by the mathematical techniques traditionally taught to the engineer. For example, the equations of motion of a simple control system may be written down and solved by, say, the method of the Laplace transform. Usually any equations used to describe the behaviour of a system are an approximation but often we are prepared to accept such approximation. In the case of a large, complex system, the sheer size and complexity of the problem may prevent our writing down equations which represent the system's behaviour, but even if we are able to write down such equations, we may not be able to solve them. Even when a system is not complex, we may not be able to write down mathematical equations which describe its behaviour with an acceptable accuracy. Since we cannot often afford to build a large system just to try it out, we often resort to the use of models. The behaviour of the model in different situations may then be found by experiment. The mathematics of fluid dynamics may be applied to the design of a boat hull, an aerofoil or a suspension bridge, only with so much approximation that we would not be prepared to gamble millions of pounds on the strength of any direct calculations. We can, however, test in wind tunnels physical models based on the calculations and have considerable confidence in the results of such tests.

A complex control system with non-linearities may not be readily solvable by classical methods. We can however, build an analogue of the system on an analogue computer and the behaviour of the analogue will (if the model is correctly designed) tell us how the real system will work. The steady-state performance of a gas turbine, a refrigerator, a nuclear reactor (the calculation of which is largely iterative) in various ambient conditions, may be modelled on a digital computer. Such a technique would be used, for example, if we wished to find the performance of a gas turbine in particular conditions of altitude, speed, temperature, etc. Running an actual test would be prohibitively expensive and manual calculation would not yield results in reasonable time.

It should be mentioned that a system does not necessarily consist of hardware. A system may be a manufacturing procedure and the term is particularly often applied to computer software, which aids management. Just as in the case of hardware, we frequently study procedures and organization on a model, usually a digital computer model, and many of the techniques used to help us to design hardware may also be used to help us to solve management problems.

It is clear that almost any engineering designer will find that he is a systems designer and important features of the 'systems' approach are:

The system has a clearly defined objective.
It will be thought successful only if a previously stated objective is met. Sub-systems and components will be successful only in so far as they contribute to the achievement of the system objective.

The system designer must co-ordinate the work of specialists.
In any given field, the designer may find that he is less knowledgeable than the men he controls and his problem becomes one of management rather than technology.

The system designer will be a model builder.
The designer's decisions, whether in the field of technology or of management, will be based on information received from models.

CHAPTER 3
Starting the Design

We have defined, briefly, the function of the designer. His first task is to decide what the customer wants. Sometimes, of course, the customer knows what he wants, sometimes he thinks he knows what he wants but is wrong, sometimes he is not at all sure of what he wants. Most very sophisticated customers (and we have mentioned B.O.A.C. and the C.E.G.B.) know precisely what they want but these are not typical and in many cases the designer finds it necessary to formulate the objectives.

Knowing what one wants is half the battle. Strange as it may seem, the customer and the designer between them frequently manage to formulate the wrong problem. What is required first is a simple statement in ordinary language of what is wanted (this statement will not necessarily contain numbers). One of the things which bedevils a statement of the objective is that the man who formulates the problem has already started to try and solve it.

An example of this difficulty was shown in a specification for a pressure switch. In fact, only after considerable expensive work had been done, was it discovered that the customer wanted a device to prevent a missile from possibly blowing up near those who had launched it. The pressure switch was asked for because the man who wrote the specification had already assumed a particular method of meeting the basic requirement would be used but this preconception prevented the proper statement of the requirements.

The second task for the designer is to convert the customer's objective into a numerically detailed design requirement. In a recent study of the design of a helicopter, there were notes on the design of the cooling system. The basic objectives could be stated as:

(i)   cool the engine,
(ii) cool the engine oil,
(iii) cool the transmission oil,
(iv) cool the engine compartment.

These objectives had to be stated and without them the designer would not know what was expected, but before proceeding with the design he had to restate the problem in more detailed, numerical terms thus:

(i)   supply air at less than 200 °F to cool the engine and engine bay (how much?)

(ii) cool 19 1b oil/min from 275°F to between 160°F and 200°F. (A load of about 54,000 B.T.U./hr.)

Although there is still some information missing, at least it is seen to be missing and, with some reservations, we can start thinking of a solution to the problem.

Thirdly, the designer must list his resources, i.e. those things which are available and which may be useful or necessary in making the hardware work. Most systems require a source of energy and it is necessary for the designer to know what sources are available. Resources could include a supply of skilled labour to operate the system, a supply of a particular material with which to build it, a river to act as a heat sink and many other factors. In the case of the helicopter cooling system, mentioned above, it was necessary to list the available heat sinks (such as fuel, the atmosphere). In fact, the designer used cold, ram air from the atmosphere to take away unwanted heat.

Fourthly, the designer must list unavoidable factors which may influence the operation of the hardware. Into this category fall such factors as the environment (will the hardware have to operate in a region of great heat, in an atmosphere of corrosive gas, in the rain, etc? will the input contain noise or any other spurious signal?), legal standards which must be met, the need to operate with unskilled labour etc. In the case of the helicopter cooling system, any equipment had to withstand the vibration which resulted from being near the helicopter's engine, government inspection standards had to be catered for, necessary overhauls had to occur at intervals not shorter than specified by the military customer.

A common, and useful, method of defining the design problem is to construct a 'black box' model of the design. 'Black box' is a term which is applied to a device when we do not know what is inside it. This certainly applies to our design at this stage.

We have a 'black box' thus:

Fig. 3.1

and we do not know what is inside. We do, however, know what we want from it and we have converted the original broad objective into detailed engineering requirements. We do know, therefore, the required output of the black box and we can improve our situation thus:

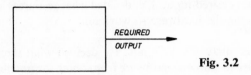

Fig. 3.2

In the case of the helicopter cooling system, the output is a required amount of cooling.

The designer also knows, or must find out, what resources he has available so that he can now bring his 'black box' up to the state of Fig. 3.3.

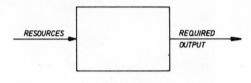

Fig. 3.3

In the case of the helicopter cooling system we have cold, ram air with which to do our cooling.

Finally, the designer has to determine those factors which will influence the operation of his product. Generally, life being what it is, there are many external factors which seem to prevent the product from working and there are some which may help it to work.

Our 'black box' has now reached the stage of Fig. 3.4.

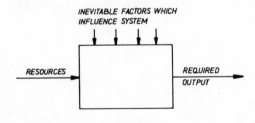

Fig. 3.4

We seem not to have progressed very far since our 'black box' is still empty, but let us list what we have done.

Assuming that we are designing some hardware, we have:

specified our required output, i.e. decided what the hardware is to do, specified our desired input, i. e. decided what resources are available to enable the hardware to work and

specified the unavoidable inputs, i.e. decided what there is that will conspire to stop our hardware from doing (or possibly sometimes help it do) its job.

This seems a reasonable and simple exercise but many designs are doomed to failure because this simple catalogue is not produced before starting to solve the design problem.

CHAPTER 4
Objectives

The manner in which the designer states his objective is very important and has a considerable influence on the system that he designs. We have already considered the possibility of not clearly stating the objective at all. It might be thought unlikely that any designer would solve a problem without first defining it, but experience shows that many designers do attempt this. Frequently this is because, at the early stage of the work, the designer has many discussions and much correspondence with the customer and the design evolves from this dialogue without a formal statement of objectives that can be referred to at a later date. Since most design involves compromise, it is inevitable that, where there has been no formally agreed statement of the problem, the designer will produce a system which will not fully satisfy the customer. The question of legally binding specifications will be discussed later, however, and the difficulty at this stage is that a designer who does not state his problem before solving it is likely to be solving the wrong problem.

We have also considered the possibility of stating the wrong problem. Usually this comes about because the designer (or customer) is already thinking of a solution before he formulates the problem and an example has already been quoted above of a case in which this actually happened. One safeguard is to use very simple English when first formulating the problem. If you are setting out to measure airspeed, prevent an engine from overheating or ensure that a gear box will not seize, then you should say so plainly and directly. With a simply defined aim, you will be able to judge from time to time whether you are being successful in your attempt to solve the problem. If the first formulation of the problem is already in some technician's jargon, then the objective will be less clear, and you will be in greater danger of limiting the possible solutions. There can be few practising designers who have not, at some time, lost a job to a competitor, not because the competitor was a better mathematician or a better scientist, but because he had more accurately judged the customer's need.

It is not always a simple matter to decide how far back to go in interpreting the customer's need. For example, if the problem is said to be to prevent an engine from overheating, we could ask ourselves whether we would be better employed in designing an engine which will operate satisfactorily at very high temperatures. In practice, much depends on the specialist skill of the designer, the specialist techniques available in the factory, and luck. If the designer has specialist knowledge of cooling systems and works for a company manufacturing heat exchangers then he would clearly set himself a different problem from the metallurgist working in the field of high-temperature materials. We must recognize however, that different methods of formulating a problem may well lead to different solutions.

Once the problem has been formulated in simple terms, the designer will probably have to turn the simply stated requirement into a numerical and technical statement. For example, if it is agreed that the designer's problem is to cool an engine, then the next question to be asked will be how much heat has to be removed; if the home handyman has decided to put up a bookshelf then he will have to ask himself how many books the shelf is to hold; if the C.E.G.B. want to supply domestic power for an area, they will have to decide how much power, in figures, this problem involves. This more formal, numerical statement, does not replace the original simple formulation of the problem because this would permit incorrect mathematics or interpretation of the basic objective to pass unnoticed.

The numerical statement of the problem must consider tolerances and 'off-design' conditions and the significance of these factors is best shown by example.

Let us consider the pressure switch of Fig. 2.1 and determine the significance of numerically stated objectives.

Requirement: the switch is required to make an electrical circuit when fuel pressure exceeds atmospheric pressure by less than 30 p.s.i. $\pm \frac{1}{4}$. Minimum fuel pressure differential is 0, maximum is 60 p.s.i.

Now, because of the danger of making a spark in fuel vapour, the actual electrical contacts must be in a sealed chamber. The design of such a micro-switch (i.e. the sealed contact assembly) is a specialist's problem so the system designer decided to use a proprietary component, selected from a catalogue. The micro-switch selected was known to require a load of 6 oz on the button to break the contacts, to have a hysteresis (between break and make) of 4 oz and to require a movement of the button of 0·006 in between break and make.

The performance tolerances required and the advertised performance of the micro-switch tell us that:

(i) when pressure increases, a pressure of 29·75 p.s.i. must produce less than 6 oz load on the micro-switch button.

(ii) When pressure increases, a pressure of 30·25 p.s.i. must produce more than 6 oz load on the micro-switch button.

(iii) When pressure decreases, a pressure of 29·75 p.s.i. must produce a load of less than 2 oz on the micro-switch button.

(iv) A travel of 0·006 in of the button must not require a pressure change of more than 0·5 p.s.i.

We have, in Fig. 2.1(i), shown only the topology of the pressure switch. Let us assume as a first design attempt that the force from the metal bellows is applied midway between the pivot and the button, then we see that:

0.5 p.s.i. variation in pressure must provide a variation in load on the button, of at least 4 oz, i.e., 0·5 p.s.i. variation in pressure must provide a variation in force from the bellows, of at least 8 oz.

The effective piston area of the bellows must therefore be at least 1 sq in. In fact, the designer knew that the effective piston area of the bellows was only about 60% of the actual face area so that a minimum bellows diameter is about 1·5 in.

Further: 0·5 p.s.i. variation in pressure must provide at least 0·006 in movement at the button, i.e. 8 oz load from the bellows must result in at least 0·003 in movement of the bellows head. That is, the spring rate of the assembly must not exceed $0·5 \ lbf/·003 \ in \doteqdot 170 \ lbf/in.$

In these calculations, we have absorbed all the permissible tolerance and we would be very optimistic if we believed that there will be no further claims on this tolerance band. We would therefore aim at a bellows diameter well in excess of 1·5 in and a spring rate of considerably less than 170 lb/in.

It is also clear that choosing a different mechanical advantage for the beam would increase the required bellows area and reduce the permissible spring rate (or vice versa). Obviously the designer will calculate the required geometry for different beam ratios and eventually select the most desirable.

The important thing to notice is that significant dimensions are being determined by the performance tolerances. The nominal make/break value of 30 p.s.i. has not yet figured in our calculations.

Let us now consider 'off-design' conditions and how they influence the design of our pressure switch. It has been stated that fuel pressure can exceed atmospheric pressure by 60 p.s.i. If we allow the button of the micro-switch to take the force from the bellows head in this case, it will be subjected to a load of about 45 lb (depending on the actual bellows diameter chosen). In fact, the micro-switch could not support such a load without fracturing so that some means of absorbing the over-run must be built into the system.

Here, design complication is introduced by an 'off-design' condition. The nominal design requirement of make/break at 30 p.s.i. has still not figured in the calculations.

Further difficulties arise because a customer's objective is seldom simple. Brunel may have regarded his objective as the cheap, comfortable transport of people and goods between any two important towns between London and Exeter. This of course, amounts to several objectives among which we can identify: cheap transport, comfortable transport, transport of people, transport of goods, transport from and to almost any town between London and the west country.

It is difficult for any designer to achieve one objective; to achieve, simultaneously, a number of objectives is practically impossible. Returning to our pressure switch, we will find that in addition to the performance requirements there will be required standards of ease of manufacture, reliability and size. These problems will be discussed in later chapters but we have already observed that the size of the system will have to be increased to reduce performance tolerances. Ultimately the designer will have to 'trade-off' size against accuracy. The reduction in volume has, in principle, a calculable cash value and so has the accuracy with which the nominal performance is achieved. Theoretically then, it is possible to achieve an optimum geometry which, although compromising on both size and accuracy, will give the maximum total value.

This situation is familiar in everyday life. If a man buys a car, he expects a good appearance, a comfortable ride, a high speed, a load-carrying capacity, reliability and many other features. One man will pay a lot for speed and not much for comfort and reliability while another may think more of reliability and comfort than of speed or appearance. Generally, with the resources available, the designer will have to sacrifice one feature for another. In attempting to find a good aerodynamic shape, say, for speed, he will lose comfort for the rear passengers.

Thus the designer in stating his objectives must:

state the basic customer's need in simple terms,

state the requirement in numerical terms which take account of tolerances and necessary 'off-design' objectives,

calculate the value of meeting an objective,

'trade-off' one objective against another and

compromise and optimize.

CHAPTER 5
Resources

Resources must be considered from several points of view. Considering first only the hardware that is designed, it must have an input. Many systems, in fact, are intended to convert an input into a desirable output; a wireless converts radio waves into entertaining sound; a power station converts coal (or some other fuel) into useful electricity; a motor car converts the energy in petrol into useful transport. Some inputs are specified when the objective is stated; a plant designed to convert grapes into wine must have grapes as one of the inputs; a wireless set designed to pick up the B.B.C.'s signals must have those signals as an input. Some inputs are, within limits, chosen by the designer; a plant designed to convert wine into brandy will require heat but the designer may choose his source of heat from any of the fuels that he has available; oil, coal, wood etc.

Generally, we may regard resources as the inputs to the system that are available for the designer to choose from. Choice is usually involved because the system need not always use the whole (or any) of every resource available. Typical of resources which the designer must consider, are sources of power, sources of heat, heat sinks and operator skill.

Resources must not be considered only in the context of operating the system. The designer is committed to produce instructions for manufacturing the system and he will therefore need to know what resources the factory has available. It will be no use designing a system which has to be manufactured using machines, techniques, raw materials or labour to which the manufacturer has no access. In short, the designer must consider the resources available to the customer when operating the system and the resources available to the manufacturer when building the system, and he must make sure that the system does not require resources that are not available. He must do more, however, than merely ensure that he does not call upon resources that are not available. Most resources are valuable and must be used sparingly if the designer is to produce a profitable system. The use of most resources, in fact, can be directly equated to the use of calculable sums of money.

As has already been discussed, the designer's job is to optimize profit. To do this, he will need to know not only the value of the objectives achieved but the cost of the resources used in achieving them. Calculations of the cost of resources consumed by the systems have always been part of the engineer's analysis. It has always been necessary to calculate, for example, the cost of the power consumed by the system, but many other factors which involve cost must also be considered. The cost of a system is a function of the cost of manufacture, the cost of operation and life, and when we consider cost from the point of view of the manufacturer we must consider:

What services has the manufacturer?
If the manufacturer is equipped mainly to build electronic equipment, it may be pointless to design a pneumatic control system.

What labour has the manufacturer?
If the manufacturer has mainly skilled machinists it may be preferable to consider a system of machined parts rather than a system of castings. If the manufacturer has largely unskilled labour operating an assembly line, the system may have to be designed for assembly line manufacture.

What experience has the manufacturer?
If the manufacturer has experience of a particular type of system then he will be able to build such a system more easily than one which is strange to his workers.

What development facilities has the manufacturer?
This is clearly related to the manufacturer's experience and services.

What is the time scale for design, development and delivery?
Clearly a very short delivery may make the job impossible without resorting to methods that are too expensive to contemplate.

Has the manufacturer any special policies which demand consideration?
For example it may be that to get into a new market, the manufacturer will be prepared to sell at a loss or at a cut price. The designer may therefore be expected to know that the permitted manufacturing price is larger than the customer's request appears to justify.

How many products are required?
The price per article may be very much less if a substantial number is required. Small numbers may dictate expensive methods of manufacture whereas a large number may justify expensive tooling. To build one motor car would involve considerable expense because of the handwork involved. To build a million motor cars would justify considerable expenditure on tools since only one millionth of the cost of the tools would be borne by each customer.

When we consider cost from the customer's point of view we have to consider yet other factors. Clearly, by reducing manufacturing costs to a minimum we are reducing the initial price to the customer, but this is only part of the cost he must bear since he has also to pay for running and maintaining the product over its useful life. In this area we have to ask ourselves the following questions:

What services has the customer?
Clearly the customer will not be interested in an air-driven system if he has only electric power available, unless the designer can justify the extra expense to the customer, of changing his plant.

What experience has the customer?
This is a question related to the services of the manufacturer and the customer will generally find it cheaper to operate a system when he has experience of systems of a similar type.

What operators has the customer?
Just as we ask 'What services and labour has the manufacturer?', the customer will generally find it cheaper to operate a system of which his employees already have experience. If this is not possible then the costs of employing and training suitable operators may be significant.

What is the delivery date?
Just as the delivery date may worry the manufacturer, it may also be significant among the customer's requirements. He may be prepared to pay for the product only if it reaches him by a given date or he may have to penalize the manufacturer, contractually, for late delivery.

What will maintenance cost?
Maintenance is one of the customer's greatest expenses and he must budget for it. Not only do repairs cost money but loss of use of a system while it is being repaired may also involve the customer in financial loss. An aeroplane with a faulty engine involves the operator in the cost of engine repairs and also in loss of revenue while the aeroplane is grounded.

The problem of costing maintenance involves us in considering the scheduled overhaul periods, the probability of unscheduled failures and the overall life of the product. These have to be considered eventually as parts of a detailed reliability analysis but it is necessary, before committing oneself to a design, to find out what the customer's maintenance problems are.

How long will the product usefully last?
The customer will be concerned to make a profit and in order to forecast his

profit through use of the product he must know how long its useful life will be. A car which costs £1000 and has a useful life of 10 years may be more economical than a car which costs £600 and has a useful life of 5 years.

Almost any engineering job means the conversion of money to a service; building the Severn Bridge meant converting money into a service; building the Concorde will convert money into a service; making wine converts money into a service. The money is what has to be paid for labour, materials, time. Sometimes we are loth to put a monetary value on resources but somehow our consumption of wealth has to be assessed. Sometimes we make these judgments without realizing it for we are prepared even to place a value on human life. If we thought human life were sacred we would never commence a major civil engineering project.

CHAPTER 6
Environment

Some of the most important factors outside our control come from the environment in which the system is situated. Obvious cases are that in designing the G.W.R., Brunel had to consider the geography of the land he was traversing; in designing the jet engine, Whittle had to consider the different altitudes at which the engine would have to operate; in designing a motor car, the designer must take into consideration that it sometimes rains.

For formal consideration we may look at:

(i)   The ambient temperature.
(ii)   The ambient pressure.
(iii) Acceleration.
(iv) Vibration.
(v)   Contaminants.
(vi) Climate.
(vii) Installation limitations.
(viii) Effect of our system on its surroundings.
(ix) Other environmental factors.

We consider below some of these factors in more detail. The examples given do not attempt to be exhaustive; no doubt every student will think of further ways in which environment can affect the performance of hardware.

Ambient Temperature (i)

If we are moving air in our system, then the colder the air the more dense it is and the more we will shift so that our system performance will vary with different temperatures. If we are moving water and we have very cold conditions, the water may freeze and our system will stop working. Consider, for example, the cooling system of a motor car. It is the possibility of a cold ambient that makes us decide to put anti-freeze in the radiator.

If we are dealing with solid materials then we may have to consider the ways in which the properties of these materials are affected by temperature. Suppose that we have an aircraft designed to operate at low speeds, then we may build it of an aluminium alloy, and it will be strong enough at all the temperatures that it is likely to encounter. If, however, the aircraft is intended to travel at speeds in excess of Mach 2, it may heat up to temperatures so high that light alloy (which gets weak at high temperatures) is not strong enough and we decide that Titanium or steel must be used for the structure. Consider the high temperatures encountered in gas turbines. These led to the development of whole ranges of alloys which would retain sufficient strength in the working ambient and indeed the gas turbine was not viable until some new alloys had been produced. Not only do metals become weak at high temperatures but their other properties change. Thus we may design a control system which depends on a spring having a known stiffness. The modulus of rigidity of a metal changes with temperature so that we may have to consider how the performance of the system changes with different ambient temperatures. Consider bought-out, electronic components of a system. Few of these work satisfactorily in temperatures much above 100° C so that a hot environment may force us to avoid the use of electronic controls. Alternatively, we may decide to use electronic controls and add a refrigerator to the system to keep the electronics cool. Many electronic systems generate heat within themselves and raise the temperature environmental to many of the components so that refrigeration or ventilation of these components may have to become part of the system design.

Sometimes human beings are part of our system and we have to complicate our system to ensure that the men are not placed in an uninhabitable environment.

Consider the pressure switch of Fig. 2.1. This pressure switch was intended to be installed in the engine bay of an aeroplane, where the temperature could range from −55° C to 200° C. This is one reason why the designer proposed to use a metal bellows instead of a fabric diaphragm, to measure pressure.

Ambient Pressure (ii)

Ambient pressure is of obvious significance to an aeroplane designer since the whole of the performance of his product is a function of ambient pressure. In many other systems however, we have pressures which are very high or very low and we have to consider how components and sub-systems will behave.

The amount of fluid a system shifts depends on the density and hence pressure of the fluid. At high pressures, components which are perfectly satisfactory at an ambient pressure of 14·7 p.s.i.a. may be crushed. The problems of designing a submarine or a watch for use underwater may largely be those of strength and of seals. At low pressures, electrical systems may cause trouble. Switches which operate well on the laboratory bench may behave very differently at very low pressure from the points of view of sparking and the deterioration of contact materials.

Acceleration (iii)

The acceleration to which a system, sub-system or component is subjected is very important in its design. Falling from a great height does not hurt until you hit the ground and that, basically, is the acceleration problem. If a part, a man, a system is subjected to a very high acceleration (or deceleration) then by Newton's law it must be subjected to a very high force. Hitting the ground after a long drop means that the body is brought to rest very quickly from a high velocity, i.e., it is subjected to a high deceleration, i.e., it is subjected to a high force—which hurts.

In a rocket which is to carry passengers, we see that the maximum acceleration that we can permit is limited by the acceleration that a man's body can stand. In much more homely systems however, we have the problem of ensuring that high 'g' loads will not prevent our product from operating. Almost any design which contains moving parts will have its performance affected by g—think of the difficulty of inventing a clock which could go to sea. The things we make are not likely to be always well treated and if our product is transportable, we may have to guard against the shocks that it will receive in transport. There is, in fact, a whole industry built around packaging for transportation.

Consider again, the pressure switch of Fig. 2.1. This is intended to be installed in an aeroplane which is expected to manoeuvre in combat at accelerations of 9 g. If the moving assembly (beam, bellows head, part of spring etc.) has an effective weight of 2 oz at the centre of the beam then an aircraft acceleration of 9 g in the direction of the axis of the bellows, will add 1·125 lbf to the force resulting from the fuel pressure. Our calculations in chapter 4 tell us that, if the effective piston area of the bellows is 1 sq in, an aircraft acceleration of 9 g will cause 1·125 p.s.i. error in switch performance, i.e., use twice the available tolerance.

Vibration (iv)

In some industries (certainly in the aircraft accessory industry), vibration is one of the most important of all design parameters. Its importance may be judged from the experience of the National Engineering Laboratory who find that neglect of fatigue is the basic cause of more than three-quarters of the failures in service which are referred to them (6.a). Of course, not all fatigue failures result from vibration—but most of them do and in any case vibration can cause failures other than through fatigue.

There are very few things which cannot be considered as possessing stiffness and mass, which means that most things can be modelled as springs and weights. Consider a spring/mass system, mounted on the engine of a motor car (the coil, the generator etc.). In its simplest terms we may regard this system as in Fig. 6.1. If m is the mass, k the stiffness of the system, x the deflection of the mass from some datum and y the deflection of the base from some datum then the force on the mass is

$$F = m\ddot{x} = k(y-x)$$

$$\text{or} \quad m\ddot{x} + kx = ky \tag{6.1}$$

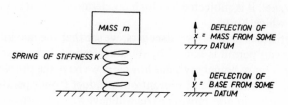

Fig. 6.1 Spring/mass system

If the base is not stationary, but is moving with an oscillatory motion such that

$$y = a \cos \omega t$$

then $m\ddot{x} + kx = ka \cos \omega t$ \qquad (6.2)

Considering only the particular solution

$$x = A \cos \omega t$$
$$\dot{x} = -A\omega \sin \omega t$$
$$\ddot{x} = -A\omega^2 \cos \omega t \tag{6.3}$$

by substitution in (6.2) we obtain

$$kA - mA\omega^2 = ka \tag{6.4}$$

$$\text{i.e., } A = \frac{ka}{k - m\omega^2} \tag{6.5}$$

$$\therefore x = \frac{ka}{k - m\omega^2} \cos \omega t \tag{6.6}$$

if $k \doteqdot m\omega^2$ (resonance) then the amplitude will be very large. If, further, ω is very large, we have

$$\ddot{x} = \frac{ka}{k - m\omega^2} \omega^2 \cos \omega t \tag{6.7}$$

and the force on the mass will be

$$m\ddot{x} = \frac{kma}{k - m\omega^2}\, \omega^2 \cos \omega t$$

with a maximum value of

$$\frac{kma \times \omega^2}{k - m\omega^2} \qquad (6.8)$$

which will be very large indeed.

It is possible, then (and in fact common) for forced oscillations to produce high forces which can both affect performance and break the components.

Consider again the pressure switch of Fig. 2.1. This switch is mounted in the power plant region of an aircraft and we see from Fig. 6.2 that it will be subjected to a vibration of up to 500 c.p.s. (6.b). We have already calculated, in chapter 4, that 100 lbf/in is a possible stiffness for the assembly and we have suggested that 2 oz could be a possible weight of the beam and associated moving parts. The system would resonate at a frequency ω, where

$$k = m\omega^2$$

$$\text{i.e., } 100 \times 12 = \frac{0 \cdot 125}{32} \times \omega^2$$

$$\text{i.e., } \qquad \omega \doteqdot 550 \text{ radians/sec}$$

$$\doteqdot 90 \text{ cycles/sec}$$

So that, as we have chosen our geometry so far, the system will resonate in service.

We have already decided that the spring rate cannot be increased beyond 170 lbf/in and to increase it to this figure would absorb all the allowable tolerance (without considering errors due to acceleration). The only possible solution to our problem (assuming the given topology) is to reduce the effective mass of the moving parts.

If we are to avoid resonance below 500 c.p.s. with a stiffness of 100 lbf/in we must have

$$m < \frac{k}{\omega^2}$$

$$\text{i.e., } \quad m < \frac{12 \times 100}{(2\pi \times 500)^2} = \frac{1200}{(1000 \times \pi)^2}$$

or the effective weight of the moving parts must be less than 0·004 lbf.

It would clearly be absurd to design the beam with an effective weight of less than 0·004 lbf unless we introduce a modification to the geometry. It would be possible to add a counterweight to the system on the other side of the pivot from the micro-switch. This mass balancing introduces problems which we cannot discuss further at this stage but suppose we can, by such means, reduce the effective weight of the moving parts to 0·002 lbf (i.e., keep resonance well above 500 c.p.s.). Fig. 6.2 shows that, at 500 c.p.s. the system would be subjected to an amplitude of 0·0008 in. In this condition, the mid-point of the beam would be subjected to an alternating force of

$$\frac{12 \times 100 \times 0·002 \times 0·0008 \times (1000 \ \pi)^2}{32 \times 12 \left(12 \times 100 - \dfrac{0·002}{32} \times (1000 \ \pi)^2 \right)}$$

$$\doteqdot 1 \ \text{lbf}$$

This would clearly be unacceptable since it would use several times the permitted tolerance. It approaches a reasonable figure, however.

The example of the pressure switch—almost the simplest example of a system that one could consider—shows that greater difficulties in design can arise from environment than from the basic requirement.

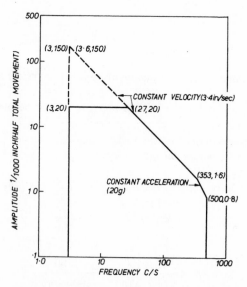

Fig. 6.2 Vibration—region of power unit. (Abstracted from B.S. 2G. 100: Part 2)

Contaminants (v)

Frequently we must design a system to work in a hostile atmosphere. Sand and dust can wreak havoc with a delicate piece of machinery. Sand can erode fast-moving components, block filters, wedge contact gaps and in many ways show up faults in equipment which worked perfectly satisfactorily in a laboratory. In chemical processes, we may be asking equipment to work in corrosive atmospheres. We may be controlling fluids which will themselves be contaminated by the use of certain alloys in our system. Even in what appear to be normal ambients, we may find contaminants that must be considered by the designer. Corrosion of steels at the sea-side is observably greater than inland.

Again the switch of Fig. 2.1 was complicated by the atmosphere. The switch was to work in fuel vapour and a spark in such an atmosphere would be very dangerous. The method of meeting this problem was to use a proprietary, sealed switch, actuated by the beam. We have seen that the use of this device dictated the minimum forces and deflections that could be used. That is, the stiffness and mass of the system resulted directly from the attempt to place the contacts in a sealed atmosphere. The stiffness and mass then presented us with apparently insoluble vibration and acceleration problems.

Climate (vi)

Climate is really a special case of a contaminating atmosphere. There are, of course, many constituents of a normal British atmosphere which could cause design difficulties in special cases but the greatest problems are caused by water. Almost always, the atmosphere contains water vapour dissolved in it. Probably, in Britain, the relative humidity (partial pressure of water vapour present/partial pressure of water vapour at saturation) is more than 50% most of the time. Slight changes of temperature result in condensation. Condensation may occur inside electrical instruments shorting out parts of the circuits. Condensation may cause water to collect in pneumatic control lines and if freezing subsequently occurs, ice prevents our control system working.

Considering again our pressure switch, we see that the inside of the bellows is open to atmospheric pressure. This will involve a small bore pipe to an orifice at the aircraft skin. We must take the elementary precaution of ensuring that the orifice at the aircraft skin does not let the rain in but even this may not be enough. Study the psychrometric chart of Fig. 6.3 and suppose the aircraft to be standing all day in an ambient temperature of 60° F and a Relative Humidity of 50% (a common atmospheric condition). Now, if the temperature falls below 42° F, water will be precipitated in the pipe and if the situation is repeated several times a considerable quantity of water could collect. If the aircraft subsequently flies in freezing conditions (which are normal above about

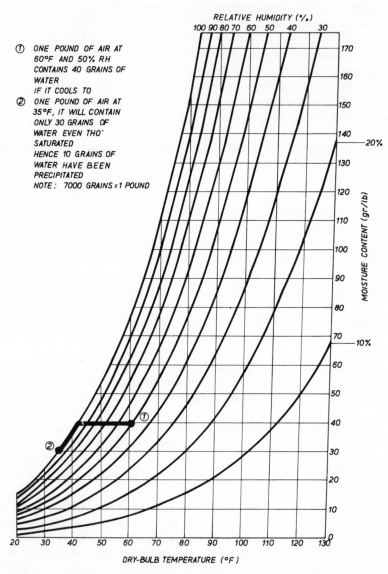

Fig. 6.3 Information abstracted from a psychrometric chart published by the Institution of Heating and Ventilating Engineers

8000 feet) there is serious danger that the pipe will be blocked by ice and the switch will not work as required. It may be possible to ensure that the pipe is self-draining but it may be necessary to keep it heated or dried chemically.

Of course, we must not forget simple and obvious effects of climate. We may have to build equipment to keep rain off (it was many years before car drivers were protected from rain as a matter of course). We may have to shelter temperature-sensitive components from the sun, flimsy components from the wind. We may have to embark on lengthy and expensive research programmes to ensure that bridges are not destroyed by wind. Not only does a bridge or a building have to withstand the high direct stresses induced by wind loads but the vortices produced may induce oscillations in the structure, which at some wind speeds are resonant. Some bridges have been destroyed by such oscillations.

Installation Limitations (vii)

Limitations imposed by available space should be obvious to the designer but nevertheless systems are still built which are too big or are built in rooms from which they cannot be extracted.

Our pressure switch had to be very small. In fact the customer originally asked that it should fit within a two-inch cube because that was the space that he had allowed in the engine bay. If we had unlimited space we could make the effective piston area of the bellows very large indeed so that small changes of pressure could create very large forces. We could then have a very stiff system to reduce the acceleration and vibration problems.

Effect on Surroundings (viii)

Sometimes the methods available to us are restricted by the effect that they would have on the system surroundings. It is mandatory nowadays that much domestic equipment, motor car equipment and the like be designed so as not to cause interference to other people's radio and television reception. When designing equipment for a ship we are clearly limited by the need to avoid interfering with the operation of the compass. In the case of large plant layouts such as atomic power stations we may have to ensure that they are not eyesores; if we are designing aeroplanes we may have to limit performance to ensure that we do not offend people with the noise of the engine.

Our pressure switch will need to be so designed that when the contacts make or break, the spark will not create a noise signal that will be picked up by the aircraft's electronic navigating equipment.

The above are merely a few examples of design situations in which the environment poses a special problem. In almost all cases, some aspect of the environment will intrude to make life difficult.

The pressure switch quoted above as an example would have been a trivial problem were it not for the environment. The environment so increased the difficulty that it was not found possible to meet the full requirement at a reasonable cost. In fact, all the problems so far considered have arisen from tolerances and environment. The nominal performance requirement has not yet figured in any calculation.

Other Constraints (ix)

Constraints may also be imposed by special circumstances such as safety requirements, legal requirements etc. Such constraints may be imposed by the customer or by a third party. The Ministry of Technology, when placing orders for certain types of equipment (e.g., aircraft flying-control systems) will lay down inspection procedures which have to be followed and which will be overseen by Ministry employees. If an engineer were to design a house, he would have to ensure that his design met the requirements of Local Authority by-laws even though those requirements were not stated explicitly by the customer. It is not possible to generalize about such constraints but in any given situation the designer must check whether any such constraints exist.

We have already seen the use of a British Standard Specification in the design of our pressure switch. Fig. 6.2 is in fact, taken from B.S. 2G 100.

CHAPTER 7
The Design Specification

Having seen that the designer must decide what he is going to do before attempting to do it, and that this involves the consideration of many factors, it is reasonable to devise a procedure that will help the designer to remember all the questions that he must pose in formulating the problem.

If we were to summarize the preceding chapters 1–6 into a check list of such questions we could obtain the following pro-forma:

1. Function (a simple statement of the objective).

2. Detailed Functional Requirements (required performance stated numerically).

3. Operating Constraints.
   3.1. Power supplies.
   3.2. Operating procedures.
   3.3. Maintenance procedures.
   3.4. Life.
   3.5. Reliability.
   3.6. Other operating constraints.

4. Manufacturing Constraints.
   4.1. Manufacturing processes available.
   4.2. Labour available.
   4.3. Development facilities available.
   4.4. Delivery programme.
   4.5. Number off.
   4.6. Permissible manufacturing cost.
   4.7. Other manufacturing constraints.

5. Environment.
   5.1. Ambient temperature.
   5.2. Ambient pressure.
   5.3. Vibration.
   5.4. Acceleration.
   5.5. Contaminants.

5.6. Climate.
5.7. Installation limitations.
5.8. Effect on other parts of the parent system.
5.9. Other environmental factors.

After completion by the designer, such a check list would be called a design specification.

Many firms insist that the designer write a complete, very detailed specification on a pro-forma which is as precise a check list as possible but many eminent designers believe that too detailed a check list is likely to restrict creativity. Probably in different circumstances both points of view are reasonable. If a firm's work is usually confined to a particular field then the designer can be fairly precise about the questions that he should ask himself since experience will have shown the important trouble areas. In such circumstances a detailed and precise pro-forma is probably desirable.

Fig. 7.1 is a design specification pro-forma that is an amalgam of several design specifications actually in use in some aircraft accessory and engine

---

Design Specification Pro-Forma

1. **Identifying Number**

2. **Issue Number**

3. **Function**
(In basic terms, what function is the article to perform when we have designed it?)

4. **Application**
(Of what system is this requirement a part?)

5. **Origin**
(By what means, when and by whom was the requirement first made known? Usually we give here a reference to a letter, visit, telephone conversation or other discussion)

6. **Customer's Specification**
(If the customer has already written a specification, its identifying number should be quoted, if the customer has not written a specification we should say so)

Fig. 7.1

7. General, Related Specifications
(If we are required to work within the framework of existing general specifications, standards or definitions or if existing documents are likely to be useful, their numbers should be quoted)

8. Safety
(Are any special safety precautions to be taken)

9. Environment
    9.1   Ambient temperatures
    9.2   Ambient pressures
    9.3   Vibration
    9.4   Acceleration
    9.5   Contaminants
    9.6   Climate
    9.7   Installation limitations
    9.8   Affect on other parts of the parent system
           (e.g. compass safe distance, radio interference)
    9.9   etc. Other environmental factors

10. Number-Off and Delivery Programme

11. Price
(Note that this may require a complex statement if prices reduce from prototypes through increasing batch sizes)

12. Functional Requirements
    12.1   Performance and acceptable tolerances (this will generally be a complex statement of the permissible range of many variables to be obtained in the presence of stated ranges of other variables)
    12.2   Life
    12.3   Unacceptable modes of failure
    12.4   Reliability
    12.5   Servicing restrictions
    12.6   etc. Any other functional requirements

13. Any Other Relevant Information
    13.1   Limitations of manufacturing facilities
    13.2   Special procedural requirements
    13.3   etc. Any other relevant information

14. Action Required
(i.e. preparation of proposal, preparation of detail drawings, manufacture of prototypes or manufacture of full production quality)

Fig. 7.1 *(continued)*

companies. It will be seen that this pro-forma asks almost exactly the questions listed above although the grouping of the questions is different. The numerical statement of the objective is part of paragraph 12; the operating constraints are not stated explicitly although 12.2 (life), 12.4 (reliability), 12.5 (servicing restrictions), all appear while other constraints (such as power) are implied by the context of the specification and paragraph 4. Manufacturing constraints are also partly listed and partly implied for, in a real situation, the designer would be familiar with the manufacturing resources available in his factory and only special resources or unusual constraints need be listed.

Paragraphs 4 (application), 5 (origin), 6 (customer's specification) and 7 (general related specifications) have not been covered by our discussion so far but are clearly lists of information which may be of help to the designer. Application (4) enables the designer to consider the parent system of which his own system is a part. This, of course, enables the designer to assess the available inputs to his own system but may also be useful if design difficulties arise which can be solved by acceptable modifications to the parent system. If, for example, the designer were unable to meet the space limitations of 9.7, he might examine the parent system to check whether it is easier to make space available than to reduce the size of his own system. Origin (5) enables the designer to check the lines of communication if a doubt arises. The customer's specification (6) is obviously useful information for, while the designer is responsible for his own specification, he would be silly not to make use of someone else's efforts to do the same job. He would also be able to see what the customer thinks he wants. General, related specifications (7) are those which constrain us, whether the customer mentions them or not. There may be legal requirements which must be met or official definitions (such as those in British Standard Specifications) of terms and units used. Such information may be of great assistance to the designer but more important still, his design may be unacceptable if it is devised without knowledge of official, legal or standard requirements.

Examples of useful or necessary related specifications are B.S. 1500 (H.M.S.O.), which gives most of the information we require to design pressure vessels and also tells us how to do the sums in a way that will satisfy the insurance companies or *The Building Regulations, 1965* (H.M.S.O.) which lists many constraints which must be observed by an architect.

Let us consider our pressure switch once more and see whether a design specification can be written and whether such a specification would be useful. We will first use the check list of page 33.

1. Function
Provide an electrical signal when fuel pressure falls below 30 p.s.i.g.

2. *Detailed Functional Requirements*
  (i)   *Fuel pressure may be any value between 0 and 60 p.s.i. above atmospheric pressure.*
  (ii) *When fuel pressure is falling, an electrical circuit is to be 'made' at 30 p.s.i.g. $\pm \frac{1}{4}$ and remain made as pressure decreases.*
  (iii) *When fuel pressure is rising, the electrical circuit is to be 'broken' at 30 p.s.i.g. $\pm \frac{1}{4}$ and remain 'broken' as pressure increases.*
  (iv) *The 'make' pressure is to be at or above 'break' pressure (note: some such statement is necessary to ensure that there is no situation in which the contacts oscillate between make and break).*
  (v)   *When 'made', the contacts must pass a current of 1 amp at 24 volts $\pm \frac{5}{6}$ (D.C.) with a contact resistance of not more than 0·05 ohms. (Note: we should really also discuss the nature of the electrical load, whether inductive etc., and a commonly used requirement is that the load to be made is such that inductance in milli henrys multiplied by the current in amps is 0·6.)*

3. *Operating Constraints*

  3.1. *Power supplies*
  (i)   *Electrical supply 24 volts $\pm \frac{5}{6}$ (D.C.).*
  (ii) *Ambient air pressure tapping available.*
  (iii) *Fuel pressure tapping available.*

  3.2. *Operating Procedures*
  The system is to be fully automatic, requiring no intervention whatsoever, during the time that it is required to operate. It must be possible to demonstrate the integrity of the system by means of a simple pre-flight check.

  3.3. *Maintenance Procedures*
  The system must not require servicing at intervals of less than 1000 hours. It will be permissible for prototype units to have shorter overhaul periods provided the overhaul period of 1000 hours is achieved within 1 year of delivering the first system.
  Any scheduled servicing must be capable of being carried out in the field, i.e. by fitters using only the normally available hand tools. If this is not possible, simple special tools may be permitted after discussion between the customer and the supplier. It is anticipated that some sort of air line (probably up to 100 p.s.i.g.) will be available to the fitters carrying out field maintenance.
  (Note: there should be more detail here but it is quite obvious that a maintenance procedure must be discussed further. It is also clear to the designer that he must give some thought to problems of easy maintenance.)

  3.4. *Life*
  (i)   *10,000 flying hours.*

(ii) A shelf life of up to 2 years must not cause the system to depart from any other requirement of this specification.

(Note 1: it is important to notice that quite high standards are required for sophisticated markets. In the aircraft industry 1000 hours would be regarded as a very short overhaul period but with a motor car component, this would be equivalent to a service every 20,000 miles or so.)

(Note 2: shelf life should never be forgotten. A lot of materials can change on the shelf.)

3.5 Reliability

A M.T.B.F. of 100,000 flying hours must be achieved during the first year of operation. (Note: the meaning of this will be discussed in later chapters.)

4. Manufacturing Constraints

4.1. Manufacturing processes available

The manufacturing organization normally makes small mechanical instruments. Only general-purpose machinery is available. No foundry is available within the factory. Limited electronic-circuit manufacture with bought-out individual electronic components is carried out within the factory. The factory already makes a range of stainless steel bellows and diaphragms.

The design should only require manufacturing procedures that are not available within the factory if it can be shown that those procedures are available from an approved sub-contractor at a price that does not reduce the profitability of the job.

4.2. Labour Available

Instrument design and development team approved by Ministry of Technology for the design of aircraft components. Prototype shop of skilled, general-purpose turners and fitters. Production department of semi-skilled, general-purpose turners and fitters and semi-skilled female assembly labour. A.I.D. approved inspection department.

(Note: a designer would not normally catalogue the resources available within his employer's factory. It is useful for the student, however, to consider the resources likely to be needed for anything he designs.)

4.3. Development Facilities Available

(i) All operating conditions can be simulated in the laboratory.

(ii) Environmental tests required by B.S. 2G 100 can be carried out provided components do not weigh more than 2 lb or occupy more than a foot cube.

4.4/5. Delivery Programme and Number off

Two prototypes to be delivered for experimental use *(in flight)* by the customer within three months of receipt of order. These two prototypes must have been proved for 100 hours life by laboratory testing. Other prototypes will be required for testing in the manufacturer's laboratory.

100 production units are to be delivered at a rate of not less than five per month, starting six months from the receipt of order.

4.6. Permissible Manufacturing Cost

Sales price has not yet been negotiated, but is expected to be of the order of £100 per switch for production systems and £500 per switch for prototypes.

4.7. Permissible Expenditure before Receipt of Order

Three companies are believed to be competing for the order. The design department is authorized to spend £500 in preparing a scheme for submission to the customer.

(Note: the whole order is worth about £10,000 with a possible profit of, say £2000. We are already committed to spend £500 regardless of the fact that there may not be a better than 30% chance of getting the order. Obviously, strict control will be required. The design problem already looks difficult and we have given ourselves three or four man weeks—at £3 per hour—to solve it and sell the solution to the customer.)

5. Environment

5.1. Ambient Temperature
$-70°C$ to $+200°C$

5.2. Ambient Pressure
1 p.s.i.a. to 16 p.s.i.a.

5.3. Vibration
B.S. 2G 100, Grade A
(Note: this corresponds to the conditions shown on Fig. 6.2.)

5.4. Acceleration
X direction $\pm$ 9 g
Y direction ±14 g
Z direction $\pm$ 9 g

5.5. Contaminants
Fuel and fuel vapour.
Lubricating oil.
(Note: we should really specify these in more detail since the exact composition of the fuel, say, could affect our choice of materials.)

Pressure and atmospheric tappings are not filtered and could give rise to small quantities of sand and dust within the system.

5.6. Climate
System could be subjected to 100% R.H. at 30°C with subsequent variation in temperature from $-5°C$ to $+70°C$.

(Note: although this is not the widest range of temperatures, this range does ensure repeated evaporation, condensation and freezing.)

5.7.   Installation Limitations

(i) Unit must be contained completely (including pipe and electrical connections) within a 2 inch cube.

(ii) Any electrical or pipe connections must conform with a commonly used standard.

(Note: this gives more scope than usual. Often all connections are dictated in advance. Here we are merely restrained from inventing special connections for which tools will not be available.)

5.8. Effect on other Systems.

The pressure switch must create no signal which has an adverse effect on other systems in the aircraft.

(Note: sparking could cause trouble to electronic equipment.)

If we now study the commercially used specification of Fig. 7.1 we will see that we could usefully have supplied the following further information:

6. Identifying Number 1001

7. Issue Number 1

8 Application

The pressure switch forms part of a safety override of the manual engine control system at altitudes above 40,000 ft.

(Note: you may find that it is worthwhile to discuss alternatives to the pressure switch to achieve the same result in the parent system.)

9. Origin

Memorandum from Sales Manager, dated 1.4.69.

(Note: now you know where to start asking questions.)

10. Customer's Specification

None supplied.

(Note 1: now you know that you must obtain the customer's agreement to your interpretation of his needs.)

(Note 2: in fact, the major engine companies usually produce very comprehensive specifications.)

11. General Related Specifications

B.S. 2G 100.

(Note: this has already been mentioned in earlier paragraphs.)

12. Unacceptable Failure

Any failure of any component in the system should result in the contacts 'making'.

13. Action Required
Prepare a scheme for submission to the customer with the object of being given the order to make the pressure switches.
(Note: this has already been implied by paragraph 4.7 but can profitably be stated explicitly.)

If the above 13 paragraphs were to be regarded as the first issue of the specification for a pressure switch, there would be some merit in rearranging the order of the paragraphs. Clearly paragraphs 6, 7, 8, 9, 10 and 13 would precede any other information.

In situations where the designer may be asked to solve problems in any field, a very precise design specification pro-forma may be regarded as too restrictive to permit reasonable innovation. Figs. 7.2 and 7.3 are the forms PAB1 and PAB2 of the PABLA (Problem Analysis By Logical Approach) system devised by Latham, Taylor and Terry of the A.W.R.E. (7a). The situation which gave rise to these forms was one in which the same designer would be expected to tackle difficult problems in quite widely differing fields. The various headings of the PAB forms have been devised to restrict the designer as little as possible while providing reminders of those areas which must be considered before he commits himself. The headings are not even intended to have precisely defined meanings which cover all occasions but they are expected to guide the designer's thoughts sufficiently to enable him to give the headings precise meanings in any particular situations. This looser approach can, nevertheless, show all the factors that must be listed in the problem statement. PAB1 (Fig. 7.2) asks that the customer's objective be stated in simple terms and then be translated into numerical, technical, engineering data. The check list (A1–A15) enables the designer to rank important factors. PAB2 (Fig. 7.3) really lists the customer's and manufacturer's resources, factors which will affect costs of operation and manufacture and environmental factors.

If we try to use PAB1 and PAB2 as the basis of a design specification for our pressure switch, we could obtain Figs. 7.4 and 7.5. These are incomplete and it would be a useful exercise for the student to complete these forms and show that all the information required of a design specification can be checked by the PABLA system.

The difference between the PABLA system and the more formal system of Fig. 7.1 becomes apparent when you attempt to solve the same problem, using both sets of forms. PABLA asks less precise questions and tends to receive less precise answers. This is not necessarily a disadvantage because too precise a formulation of the problem might restrict the creativity of the designer. Generally a company working in a particular area will use a fairly precise format for a design specification where the jobbing designer will prefer the loose constraints of PABLA.

It is not difficult then, for the designer to adopt a systematic procedure when formulating his problem. It is reasonable to devise a check list to help this work and the form that this check list must take, provided that it is comprehensive, may be tailored to the design situation and the designer's tastes. So far however, we have considered the design specification entirely as an aid to the designer. In fact, the customer must also be considered because, eventually, a version of the design specification will form the basis of a contract between the designer (or his employer) and the customer. The broad statement which emerges from an informal system, while it may assist the designer, is of little help in any commercial transaction with the customer.

The designer is selling his skill and the manufacturer is selling his products but usually the price is agreed and the order placed before the designer has completely demonstrated his ability or the manufacturer has made his wares. In this situation, the design specification is the only real statement of what the customer expects to get and what the manufacturer contracts to supply. The design specification is the only yardstick by which later achievement (or lack of it) can be measured.

The form of Fig. 7.1 has been designed to act as a basis for a contractual agreement and for this reason it carries identification, issue number, reference to the origin of the requirement and reference to any formal statement of requirement that has been made by the customer. This document is generally part of the manufacturer's formal drawing system so that the records which are kept of the specification and the procedure for making changes to it are covered by a formal drawing modification procedure. Orders placed may then refer to a particular specification at a particular issue number.

Where the designer prefers, for his own use, a less formal approach, it is still necessary for him to produce a detailed specification to form the basis of the financial agreement with the customer. It may be reasonable then for an informal system, controlled only by the designer, to exist side by side with a more formally documented system.

The necessity of a formal specification cannot be overemphasized. However pleasant the relationship between the customer and the designer may be at the early stages, it is certain that, sooner or later, there will be a disagreement about whether the customer's requirement has been met and the only way that such a disagreement can be settled is by reference to the design speci-fication quoted on the order.

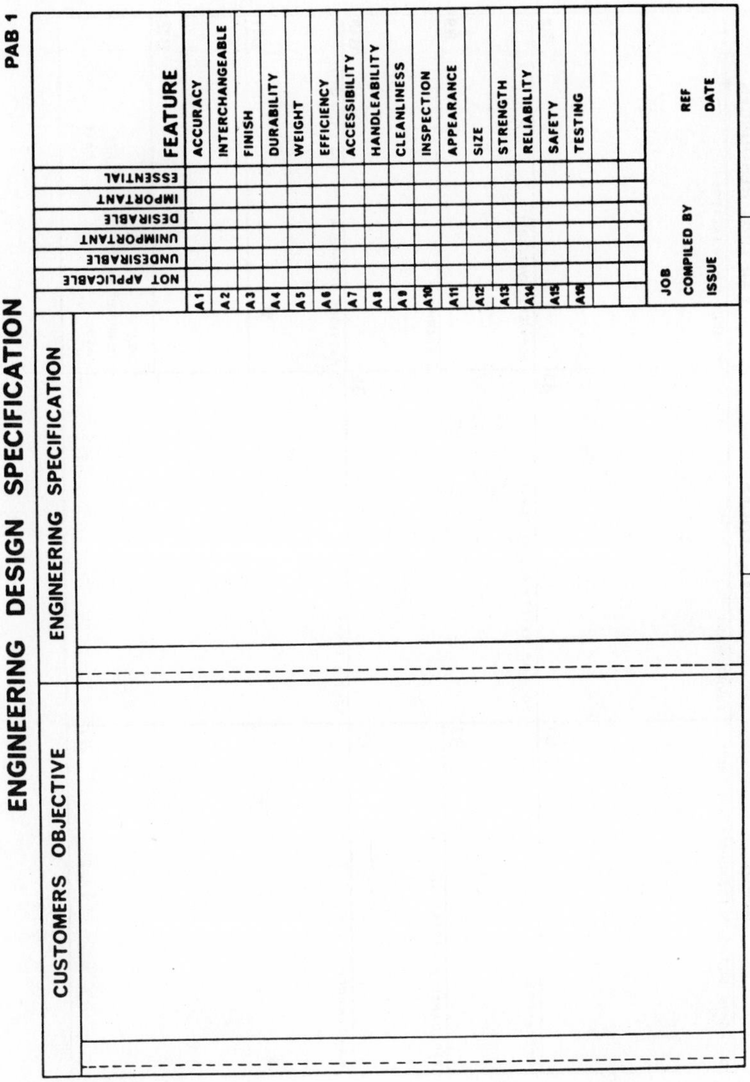

ENGINEERING DESIGN SPECIFICATION

PAB 1

FEATURE

ESSENTIAL
IMPORTANT
DESIRABLE
UNIMPORTANT
UNDESIRABLE
NOT APPLICABLE

ACCURACY — A1
INTERCHANGEABLE — A2
FINISH — A3
DURABILITY — A4
WEIGHT — A5
EFFICIENCY — A6
ACCESSIBILITY — A7
HANDLEABILITY — A8
CLEANLINESS — A9
INSPECTION — A10
APPEARANCE — A11
SIZE — A12
STRENGTH — A13
RELIABILITY — A14
SAFETY — A15
TESTING — A16

CUSTOMERS OBJECTIVE

ENGINEERING SPECIFICATION

JOB
COMPILED BY
ISSUE

REF
DATE

Fig. 7.2

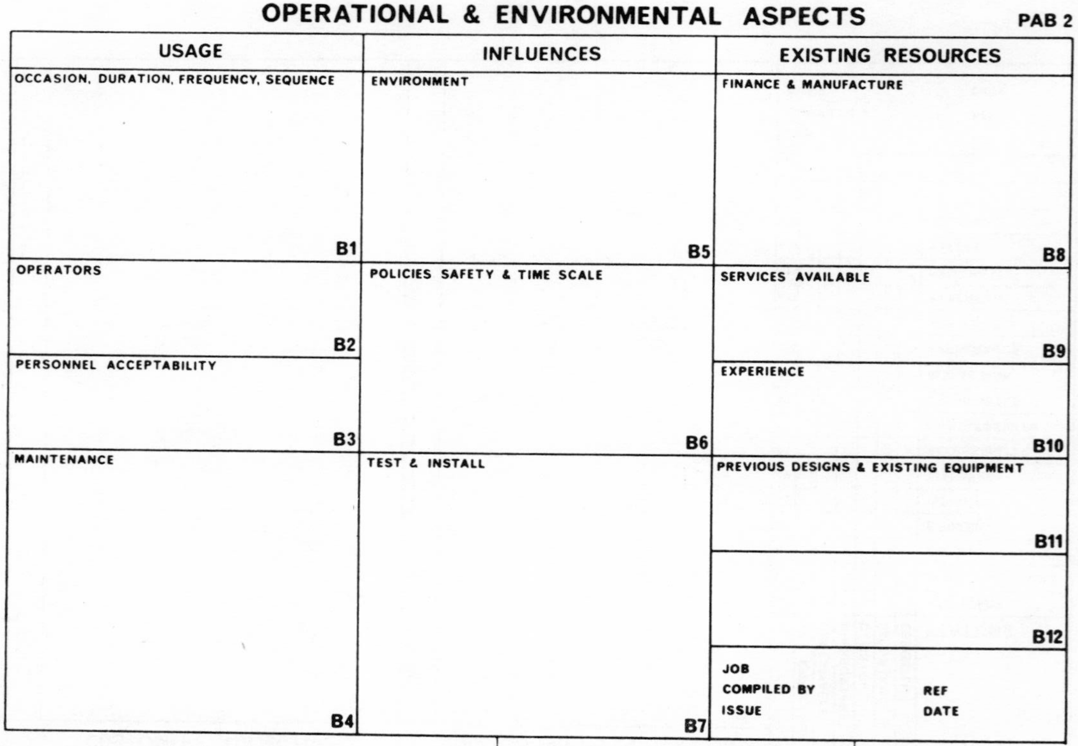

OPERATIONAL & ENVIRONMENTAL ASPECTS PAB 2

USAGE	INFLUENCES	EXISTING RESOURCES
OCCASION, DURATION, FREQUENCY, SEQUENCE	ENVIRONMENT	FINANCE & MANUFACTURE
		B8
B1		
OPERATORS	POLICIES SAFETY & TIME SCALE	SERVICES AVAILABLE
		B9
B2	B5	
PERSONNEL ACCEPTABILITY		EXPERIENCE
		B10
B3		
MAINTENANCE	TEST & INSTALL	PREVIOUS DESIGNS & EXISTING EQUIPMENT
		B11
	B6	
		B12
		JOB
		COMPILED BY REF
		ISSUE DATE
B4	B7	

Fig. 7.3

ENGINEERING DESIGN SPECIFICATION

PAB 1

FEATURE	ESSENTIAL	IMPORTANT	DESIRABLE	UNIMPORTANT	UNDESIRABLE	NOT APPLICABLE	
ACCURACY	✓						A1
INTERCHANGEABLE	✓						A2
FINISH		✓					A3
DURABILITY	✓						A4
WEIGHT	✓						A5
EFFICIENCY	✓						A6
ACCESSIBILITY			✓				A7
HANDLEABILITY		✓					A8
CLEANLINESS		✓					A9
INSPECTION		✓					A10
APPEARANCE				✓			A11
SIZE	✓						A12
STRENGTH							A13
RELIABILITY	✓						A14
SAFETY	✓						A15
TESTING	✓						A16

JOB X 1001

COMPILED BY D.J.L.   REF

ISSUE   DATE 20/10/69

CUSTOMERS OBJECTIVE

Provide an electrical signal when fuel pressure falls below 30 p.s.i.g.

ENGINEERING SPECIFICATION

(i) Full pressure between 0 & 60 p.s.i.g.

(ii) Falling pressure must make contact at 30 p.s.i.g ± 4 p.s.i. and the contact must remain closed as pressure falls.

(iii) Rising fuel pressure must break contact at 30 p.s.i.g. ± 4 and the contact must remain open as pressure rises.

(iv) Make pressure must be at or below break pressure.

(v) When made, contact must pass a current of 1 amp. at 24 volts D.C. with a contact resistance less than 0.05 Ω

Fig. 7.4

OPERATIONAL & ENVIRONMENTAL ASPECTS

PAB 2

USAGE	INFLUENCES	EXISTING RESOURCES
OCCASION, DURATION, FREQUENCY SEQUENCE Once per flight ~ say once in 3 flying hours. Life 10,000 hours. B1	ENVIRONMENT BS 2G 100 Vibration grade A Acceleration X ±9G Y 14G Z ±9G Temperature −55°C to +200°C Fuel Vapour B5	FINANCE & MANUFACTURE Limit expenditure to £500 for proposal. Cost £400 each for prototype £100 each in production B8
OPERATORS Maintenance by R.A.F. technical personnel B2	POLICIES, SAFETY & TIME SCALE MTBF 100,000 hours Delivery 2 prototypes 3 months from order 100 production at 5/month after 6 months from receipt of order. B6	SERVICES AVAILABLE B9
PERSONNEL ACCEPTABILITY B3		EXPERIENCE B10
MAINTENANCE Not more often than once per 1000 flying hours. Service to be carried out in field or with simple equipment. Works air line available. B4	TEST & INSTALL Test before each flight B7	PREVIOUS DESIGNS & EXISTING EQUIPMENT B11 Proposal only at present B12 JOB X1001 COMPILED BY D.J.L. REF ISSUE 1 DATE 20/10/69

Fig. 7.5

Exercises and Subjects for Discussion on Part 1

Exercise 1 is normally used as a basis for an afternoon's discussion with students. The questions are not set as such but are fed into the discussions as necessary and frequently it will be found that the course of the debate makes it unnecessary for the leader to inject all the questions. The idea of the discussion is to get the students to argue about the ideas presented in the preceding chapters, in contexts which are familiar and which require no special knowledge of Mechanics.

Exercise 2 enables the students to crystallize the results of the above discussion. Usually groups of 3 or 4 will work for 2 or 3 hours on the exercise. A student who works in a small group usually seems to learn more than one who works alone.

Exercises 3 and 4 are based on examination questions and can be set to be done in about $\frac{3}{4}$ hour. The student is expected to do no more than list, and briefly discuss, the main points in his argument.

Exercises 5, 6, 7 and 8 are straightforward exercises in writing design specifications. Exercises of this nature can take an hour or two if we are not too fussy about the accuracy of information, or several hours spread over a week if the students are asked to search for data. Exercise 5 shows a worked answer that resulted from an afternoon's work by several students.

Exercises 9 and 10 are slightly more advanced exercises in design specification writing. These require some research and are really starting to require specialist knowledge.

One or two exercises, at least, should be studied in some depth so that the student will be able to use a design specification that he has prepared as the basis for exercises to be done after reading later chapters.

1. Mr. Jones says he wants a house and a car.
What does he really want?
Would he be happy in a flat or a caravan?
Would he settle for a van instead of a car?
Would a motorcyle do?
Would he like free travel on the railway?

What would we like to know about Mr. Jones before we can say what he wants? His job? His age? His salary? His hobbies? His holiday plans?

When we know what Mr. Jones wants, can we start designing it?

What can Mr. Jones afford?
Where is the house to be and why is this significant?

Where is Mr. Jones going to use his car and how much will he pay for it?

Why will Mr. Jones be prepared to pay more per mile for travelling by car than by train?

What features of the house may be traded against other features and what are the relative values?

What features of the car may be traded against other features?

What materials shall we use for the house?
Would we use the same materials if Mr. Jones lived in Canada?

What environmental factors are important in the design of the house?

What environmental factors are important in the design of the car?

List some of the sub-systems of which Mr. Jones' house is the parent system.

List some of the sub-systems of which Mr. Jones' car is the parent system.

What is the environment inside the bonnet of Mr. Jones' car?

Which of the sub-systems in Mr. Jones' car will be affected by the environment under the bonnet?

2. Devise a form of design specification which
   (i) will be useful to an architect designing houses
   (ii) will be useful to an engineer designing cars.

   Use your pro-forma to write a specification for
   (i) Mr. Jones' house
   (ii) Mr. Jones's car.

3. In a small, privately-owned coal mine, the miners have broken through to some old, disused workings and water is running from the old workings into the mine and flooding the working area. The mine manager has asked for a pump to get rid of the water and make the working area accessible again.

   Would we, in fact, need a pump?
   Assuming a pump to be desirable, what significant environmental factors might be important in the design of the pump?

   How does money come into the exercise?

   How much does it cost us if the coal face is flooded?

4. What would be your objectives and how would you turn them into precise, engineering requirements, if you were asked:

(i) to create a habitable environment for the 2-man crew of a satellite which will be in orbit for 10 days?

(ii) to create a habitable environment for the fare-paying passengers of a supersonic transport aeroplane?

(iii) to create a habitable environment for the passengers of a double-decker bus in the Swansea town services?

5. Your boss has recently had a telephone conversation with the Chief Constable of Glamorgan who has stated his need for a device with which a policeman can test the state of the brakes of any car, stopped at random, anywhere in the county. The tests are to be done at realistic car speeds. The Chief Constable is thinking of purchasing two such devices for trial purposes and if these trials are successful he would order a further 50. Your boss asked the Chief Constable if there were prospects of sales outside Glamorgan and as a result of the discussion he feels that success in Glamorgan could be followed by nationwide sales.

Write a design specification for the job.

This particular example is entirely artificial and is written about a situation reasonably familar to a student. It is an example that is typical of the way many jobs start and quite clearly there is insufficient information to start a design. The design specification may, however, be drafted so that we realize what gaps there are in our required knowledge and also so that limited design work can commence. Under no circumstances should design work start until a design specification is written. It is significant that when the problem was first set, the students concerned were also asked to suggest ways of meeting the requirement. A large proportion of the students commenced design work without first defining the problem—a situation which is paralleled in many commercial design offices where the designer will frequently have a design scheme on his board before the problem is defined.

A possible first draft answer to the above problem which resulted from the deliberation of a small group of students was as follows. Comments are offered in parenthesis.

1. **Design Specification no.:** D.J.L. 1.

2. **Issue no.:** 1.

3. **Function:** Road Vehicle Brake Tester.

4. **Application:** Vehicle Testing Programme of County Police Force.

5. **Origin:** Telecon between Managing Director and Chief Constable of Glamorgan on 21.2.67.

6. **Customer's Specification:** None available.

(This is true at the time of writing this issue of the specification but it is possible that there is a specification in some form and we should check this point. The entry 5 tells us where to start making such enquiries.)

7. General Related Specifications. Legal requirements for safety of road vehicles. (The student would not be expected to know these requirements by name but should realize that some official requirements will exist and the designer should make enquiries so that they can be quoted in detail in the next issue of the specification).

8. Safety. Conditions of test and use of the equipment must not increase the danger to the car being tested, its occupants or the health or property of other road users.

9. Environment

  9.1 Ambient Temperature: for use   0° to 25°C
                   for survival   −20° to 45°C

  9.2 Ambient Pressure: 14·7 p.s.i.a. ± 2.

  9.3 Vibration: If used inside a vehicle the system must operate in a vibration ambient of 16 c.p.s. to 70 c.p.s. (corresponding to engine r.p.m. of between 1000 and 4000). The amplitude of such vibrations is not known at the time of writing issue 1 of the specification but is to be ascertained. For the purpose of initial calculations it will be assumed that the amplitude is $\pm$ 0·01 in. It may be necessary to consider higher frequencies (engine speed × No. of cylinders) of very small amplitudes.

  There may also be vibrations caused by uneven road surfaces and we shall consider these to be $\pm$ 1 in at 30 to 100 c.p.s.

  9.4. Acceleration: The system will operate with a stated accuracy at accelerations between ± 2 g. The system must not be deranged by accelerations between ± 4 g.

  9.5. Contaminants: The system must operate satisfactorily in the extremes of sand, dust and salt-spray likely to occur on or near roads in the British Isles. (Check whether standards exist.)

  9.6. Climate: The system must survive and operate satisfactorily in extremes of humidity likely to obtain anywhere in the British Isles and also in dry weather and rainy weather. (Check whether standard rain tests are laid down.)

  9.7. Installation Limitations: The system must be easily transportable and usable by one policeman with a light car. The use of the system must require no modification to the car being tested.

  9.8. Other Limitations and Relevant Factors:

  9.8.1. The system must be such that no motorist will be detained for more than 8 minutes.

  9.8.2. The output of the system must be suitable for use as evidence in the event of prosecution of the motorist for having inadequate brakes.

10. Number-Off and Delivery Programme

  10.1. For development and prototype test: 5

  10.2. Initial production batch: 50

  10.3. Ultimate production sales : 1000

  10.4. Delivery programme is yet to be agreed but for initial calculations, assume delivery of prototypes to customer to be 3 months after receipt of order. This will require the manufacture of prototypes for test and development within 6 weeks of receipt of order. Assume too that a first production order for 50 will be met by delivery of the full number within 3 months of receipt of order.

11. Price. To be agreed with prospective customer. Target price for purpose of initial consideration:

Prototype: £400 each
Limited production: £50 each
Production: £25 each

(This is, of course, a pure guess and some research will be required to find the price acceptable to the customer and the prices of competitive instruments in the same, or in related, fields.)

12. Functional Requirements

12.1. Performance: The system will record whether the brakes of a car are capable of producing a deceleration of 0.6 ± 0.1. (This is a guess at this stage, based on immediately available information, such as the Highway Code; it is assumed to be of the right order so that initial calculations could be made and discussions held with the customer.)

The test may be conducted at any initial speed between 60 m.p.h. and 20 m.p.h.

Note: The minimum acceptable braking has been assumed to be 0.75 g. On the worst tolerance we will be sure that our reading is $0.6 \pm 0.1 = 0.7$ g, i.e. we are sure that it is marginally less than 0.75 g. This figure is to be checked in discussions with the Glamorgan Police.

12.2. Life: 5 years.

12.3. Failure: No particular mode of failure is desirable but failure must be immediately apparent: see 12.4.1.

12.4. Reliability:

12.4.1. Accuracy to be easily checkable immediately before and immediately after a test in the field.

12.4.2. Reliability to be such that there will not be more than one failure per 10,000 tests.

13. Any Other Relevant Information

It is to be determined whether Police Forces have a special procurement or servicing procedures which must be followed.

14. Action Required

Preparation of a brochure for submission to the Chief Constable of Glamorgan. This brochure must be submitted within 3 weeks of the date of issue 1 of this specification. The brochure must show how the requirements of the latest issue of the specification, before the date of the brochure, will be met.

In drafting the above specification, no attempt was made to finalize numbers. There are two reasons for this:

(i) Writing the specification is the first task which must be tackled; this discipline will tell us where we lack information. It may frequently be possible to assume information which, while not accurate, will be sufficiently of the right order to enable us to consider possible solutions to the problem. This is true of the industrial situation.

(ii) The problem was written as an undergraduate exercise and we would gauge a student's understanding of specification requirements without his having to undertake the actual research necessary to obtain precise figures.

6. Write a design specification for the passenger seat of a supersonic passenger transport aeroplane.

7. Write a design specification for a pressure switch to indicate whether oil pressure is adequate in a cheap family motor car.

8. Write a design specification for a device for the day-to-day cleaning of a 3-bedroomed house.

9. During the war, aeroplanes frequently flew at altitudes up to 20,000 ft. The Germans built and successfully flew a reconnaissance aircraft at over 30,000 ft and the R.A.F. were unable to fly at the altitudes necessary to attack this aircraft. It was decided that some Spitfires would be modified to enable them to fly sufficiently high to attack the German reconnaissance aircraft and among the jobs necessary to this exercise was the task of devising the method of pressurizing the fuel. The fuel was necessarily pressurized to prevent evaporation. This task was given to a small engineering company and the requirement met in a very short time.

Draft a specification for such a fuel tank pressurization system.

This example has an easily defined performance requirement but it has, in addition, complex environmental factors which must be defined.

10. Discussions with horticulturalists, meat traders and egg producers suggest that some fruit, meat and eggs could be better stored if they were chilled in such a way as to prevent evaporation. The cold rooms used at present have very dry atmospheres, causing loss of weight (and it is believed loss of quality) through evaporation.

This problem was in fact tackled some years ago by a commercial firm without success. No proper specification was written therefore no real analysis of the cost and profits involved in meeting the requirement was ever made. It is quite probable that a systematic approach to understanding the requirements would have led to successful designs.

Draft a specification. This exercise is included because it is an example of a job in which there is difficulty in formulating the required performance of the design. Most of the blanks in the design specification pro-forma can be filled in without great difficulty with the exception of the function and the detailed functional requirements. What, in engineering terms, is required? Is it that the refrigerator must condense water in smaller particle sizes than do current refrigerators? If so, what particle size? Is it merely a question of temperature and if so are the different requirements compatible?

In fact, the attempt to write the specification should in real life lead to the formulation of a research programme and not directly to an engineering design scheme.

Bibliography to Part 1

2(a) W. Gosling, *The Design of Engineering Systems*, Heywood, London, 1962.
6(a) *The Feilden Report*, Department of Scientific and Industrial Research. Report of a Committee appointed by the Council for Scientific and Industrial Research to consider the present standing of Mechanical Engineering Design, London, H.M.S.O., 1963.
6(b) British Standard Specification, B.S. 2G 100, part 2, 1962.*
7(a) R. Latham, H. Taylor and G. Terry, *The Problem Analysis by Logical Approach System*, Engineering Materials and Design Conference, London, 1965.

Further Reading
H. Chestnut, *System Engineering Tools*, Wiley, New York, 1965.
T. Woodson, *Introduction to Engineering Design*, McGraw-Hill, New York, 1966.

* B.S. 2G 100 is progressively being revised and updated to 3G 100. In particular, the vibration requirements quoted in the text have been deleted from the British Standard. B.S. 3G 100: Part 2: Section 3: Sub-section 3.1, now applies although some conditions are still under consideration.

Part 2
FEASIBLE SOLUTIONS

Part 2 is largely concerned with finding answers to the problems posed by the design specifications of Part 1. Chapter 8 attempts to put this problem-solving into its proper commercial context, and chapter 9 discusses methods of creating solutions. Creativity is a fashionable study at present, although it is difficult to say whether all the research being undertaken will eventually increase the number of good designs. Brainstorming and similar procedures can be seen to work when the problems are small but whether the methods have intrinsic merit or whether they merely serve to open the mind of the innovator, is not known. Students benefit from taking part in creative sessions, seeming to learn objectivity and also to acquire a more creative (or open-minded) approach in that they really seem to produce better and simpler solutions to problems after a session or two. Before taking part in group creating sessions students tend to be timid, wanting to reproduce familiar designs, copy the work of or wait for a lead from older men. Often a good brainstorming session will convince a student that he has as much right to his opinion as his seniors. Once again lectures and reading have little meaning unless backed by discussions and practical work.

Osborn, 1957, invented Brainstorming and has written about it. Gordon invented Synectics and has written about it (9b). The average undergraduate cannot, however, afford to spend a lot of time reading about creativity, which must find its proper place among other disciplines—there is perhaps a hint of excess in the idea of half-a-dozen academics spending a day or two joking into a tape recorder in the hope of inventing a tin-opener. Alger and Hayes, 1964, have produced an excellent small book which summarizes Brainstorming and Synectics among other creative aids. The *Design Method*, edited by Gregory, 1966, contains a number of papers (e.g. one by Broadbent), summarizing creative methods.

Chapter 10 does, to a limited extent, discuss methods of selecting the best solution from several available. Mention is made of Marples' (10a) work on decision-making in design. A further excellent text on decision-making is by Starr, 1963. The main purpose of chapter 10, however, is to justify and list the topics considered in more detail in subsequent chapters.

CHAPTER 8
Getting the order

So far, we have assumed that the customer has a problem, that we have defined that problem, listed the constraints which are imposed on us and the resources which are available to us. We have not yet attempted to solve the problem that we have formulated and indeed we do not know that a solution exists.

We must now convince ourselves that we can solve the problem and that it would be worthwhile for us to do so.

More important still, we have to convince the customer that we can solve the problem and that our solution is worthwhile to him.

Consider that, in the normal commercial world, our designer (or his employer) has, so far, been working for nothing. He will continue to do so for some time for he has not yet received an order and it is not usual for a customer to pay for work which the designer does in the hope of getting an order. The next task therefore is to convince the customer that he should place an order for our work. Until he places an order we continue to work for nothing and it is important that we spend as little of our own money as possible. Nevertheless the money which a designer (or his employer) spends to get an order may be considerable for not only does he have to do enough work to convince the customer but he is almost certainly doing so in competition with other designers.

The most modest problem will involve labour costing, say £2·50 an hour while, if experimental or research work is required, costs per hour may be much higher. To obtain a large construction job, the civil engineer may have to spend a sizeable fraction of a million pounds; to get the order for an aeroplane engine, the engine constructor may have to spend hundreds of thousands of pounds on design studies, research and tests; to get the order for a power station, the design team may have to spend hundreds of thousands of pounds on computation, model studies and research. Remember too, that because there is competition in every field, we may have to spend this money with a comparatively low chance of recovering it.

In most sophisticated branches of engineering, the designer will prepare a brochure for the customer and this brochure will outline the designer's proposals and arguments. Many rude things have been said about the 'art of brochuremanship' but if your bread and butter depends on getting orders then it is only common sense to do everything you can to get those orders.

If we consider the pressure switch for which we wrote a design specification in chapter 7 we will realize that some of the problems of salesmanship have already appeared and it is worth looking at these problems again.

Paragraph 10, p. 40, stated that the customer had not supplied a specification of his own. Whatever else the brochure contains, it must clearly contain a copy of our design specification so that the customer knows what problem we are solving and can approve our statement of it or can correct any misunderstanding we may have.

Permissible expenditure before receipt of order, paragraph 4.7, p. 39, stated that the design department was authorized to spend £500 in preparing a sales brochure. When we realize that two men, working for a fortnight together with a visit or two to the customer, may easily cost £500, we see that considerable discipline is required to produce a brochure likely to get an order. We see too that the Sales Manager of our factory has a very difficult task in assessing how much he can venture in the hopes of getting an order. With an anticipated profit of only £2000 if the order is obtained, £500 looks to be a high amount to risk.

Permissible manufacturing cost, paragraph 4.6, p. 39, suggested a selling price of £100 per system. The above argument however, suggests that we are investing £500 in a risky, low-return project. We have already seen that the problem is going to be difficult (we do not know how to meet the vibration conditions) so we must ask ourselves whether we really want the job at all. Probably some preliminary discussion with the customer is necessary to ensure that we really do have to meet the vibration conditions. If we do, then perhaps a higher selling price can be seen to be reasonable. Perhaps we can accept the unit selling price but negotiate a separate contract to pay for the development necessary to solve the environment problem. We are obviously required to solve a management (or, at least, a financial) problem before we put a clean sheet of paper on the drawing board.

This exercise shows us that:

(i) There must be an agreed problem and the solution of this problem must have a value to the customer. Normally the customer would be expected to know how much it is worth to him to have his problem solved but a good designer should be able to advise in this commercial assessment.

(ii) The designer must decide how much money should be risked in the hope of getting an order. In making this assessment he will have to consider the expected profit if the order is obtained, together with the chance of getting the order. This is a very nice judgment for, generally, the more the designer spends in trying to get the order, the more he raises the chance of getting the order, but the more he reduces possible profit.

When the problem is stated and the extent of pre-order design work is stated, it is then necessary:

(iii) To think of a way of solving the problem and convince, firstly himself and then the customer, that the solution is physically realizable.

(iv) To convince the customer that the proposed solution is possible with the resources available.

(v) To convince the customer that the proposed solution can be made available when the customer wants it.

(vi) To convince the customer that the proposed solution will cost him less than those of the designer's competitors.

The statements of these six arguments really provide the brochure (although it is unlikely that the results of (ii), above, will be communicated directly to the customer, however necessary they are to plan the work involved). None of these cases may be argued cynically because, ultimately, the designer's career will depend on his ability to keep promises, not on his ability to make them.

Of the above six arguments, (i) and (ii) have already been discussed while (iv), (v) and (vi) are problems of analysis. Our immediate concern is (iii), the generation and selection of a solution to the problem. Since we will wish eventually to concentrate our efforts on one solution to the problem and since we will want this solution to be a good one, we must:

generate possible solutions and

reject those that will not work. This means those that will not work because natural laws prevent it, those that will not work because they demand more resources than we have available and those that will not work because we cannot deliver on time.

Of those solutions which still remain we have to:

select the best; i.e. that which is attractive enough to the customer while making a profit for the manufacturer.

We may consider the process as one of sifting, using as coarse a mesh sieve as possible. We eliminate unsatisfactory systems before too much money is spent on their design and then repeat the exercise with a finer mesh. Ultimately, if there are several systems that are apparently suitable or if only one possible system has parameters which may be varied we must select the system which optimizes some desired objective (probably that which gives most value). Having chosen a solution to the customer's problem for reasons which satisfy us, it is necessary to propose that system with its justification, in the hopes of getting an order.

In Part I we considered, in detail, how to state the problem; we have not yet thought of ways of solving the problem. The problem statement probably decided how much we may spend on our brochure and typically, this decision was made without reference to the amount of work the designer would like to do. We have however, derived a useful sequence of actions.

The procedure suggested on p. 58 will be developed as a logical procedure for a designer. In this, we have assumed that the designer has been offered a problem which the customer requires solved. This is an acceptable approximation to most design situations. There are variants of this situation however and the nearest is that where the designer, because of his specialist knowledge, shows the customer a requirement which can be met by a new design. This does not invalidate our model but adds salesmanship and knowledge of the market to the other functions of the designer.

There are also situations in which a system has been invented without consideration of the market. There is something illogical about inventing an answer without a problem and the engineer need rarely find himself in this position although the firm's management has to decide how much money to spend on research, which really means spending money solving problems which are not necessarily commercial. Clearly there are some manufacturing organizations which could not survive without research but it is difficult to see what logic can be invoked to decide the percentage of a firm's profit that will be re-invested in research. Often the amount of money invested in research is determined arbitrarily within a figure which the firm is rich enough to write off.

Similarly, money spent on a market survey runs the risk of having to be written off completely for there is no guarantee that a market survey will find a market or be right in its predictions if it does. A marketing budget therefore is likely to be fixed arbitrarily within a figure which the company is prepared to write off.

Consider as an example a company making pneumatically-operated control valves for the chemical industry. The company may investigate the market for pneumatic control systems to operate its own valves. Once entered into such a field, the firm might invest money in research into pneumatic logic, in the expectation that the knowledge acquired would eventually be saleable. In such a case, a senior designer would expect to help with both the market survey and the research, because he has knowledge which is useful to both activities. The real design work, however, starts when the market survey has defined the problem and its value and the research merely provides a little more knowledge of possible solutions to the problem.

If we consider that the designer's job really starts when marketing has shown a customer need or a possible use for the result of some research then the procedure for providing a solution as summarized above will be a reasonable one to follow.

CHAPTER 9
Creating Possible Solutions

When we have defined our problem we have to think up one or more ways in which it may be solved. The process of inventing a solution to a problem is not understood and it is not possible to lay down rules by which possible solutions may be generated. The Greek geometers were aware of a similar difficulty in that a theorem could be proved only when its truth had already been appreciated—and what led a man to invent the theorem? As in the case of the Greek geometer, the engineer must find a possible solution to his problem before he can analyse that solution and demonstrate that it is a good, or even feasible solution. The designer has many tools of analysis at his disposal but is unable to use them until he has thought of something to analyse. This problem of finding solutions (or possible solutions) has been called heuristics but recognition of the problem has done little to help us. There are however some methods which in a limited way help us; for example:

Prior Systems

A designer who has much experience will be aware of systems, already in existence, which will solve or nearly solve, his problem. Students are often criticized for their lack of ability to innovate worthwhile designs. Much of this difficulty arises from the fact that students have very little experience and have, consequently, an awareness of only a small number of systems which may be repeated or copied. Some people believe that designers mature later than, say, mathematicians, and that this is largely because the acquiring of experience is so necessary a part of the designer's education.

If we look at our pressure switch we will realize that a designer asked to design such a system almost certainly works for a company which has made similar systems before. In such a company, the ideas of Fig. 2.1 would be commonplace. This does not, of course, mean that the severe vibration problem has already been solved, but at least the firm will have available a lot of knowledge in this field, derived from considering earlier, similar problems.

Prior Systems in Nature

Many design problems which face an engineer have already been solved by nature. Most structural problems may be seen in plants, many mechanisms exist in animals; the eye, the ear and many other organs contain elements which may usefully be copied in certain cases. It is even said that Brunel (senior) was inspired, in his creation of a Thames tunnel, by the activity of a woodworm.

Analogues

Even where direct solutions to our problems do not exist, either in art or in nature, there may nevertheless be an analogous problem which has been solved and from which we may draw inspiration. It is apparent that many systems (and particularly linear systems) work in analogous ways. Pneumatic controls, electronic controls, electrical networks, spring/mass systems, economic systems and many others may be represented by similar equations of motion and the components of one system may be direct analogues of components in another. Although such analogues are more usually considered as aids to analysis of a system already designed, they may also be used to create a system. Knowing that the performance of a pneumatic logic element may be analogous to that of a transistor, that a bellows may be given some properties of an electrical inductance, etc., then a completely new pneumatic system can be built up by analogy with an existing electrical system.

One large class of design problems is to automate processes which are already carried out manually. In such a case one can generate at least one solution by copying, mechanically, each of the actions taken in the manual process.

Many processes, currently carried out by humans, involve very little actual movement but may require calculation or, at least, thought. In such a case, we may generate a solution to our design problem rather as if we were writing a computer program, once again copying as nearly as possible the actions and processes currently being performed by humans.

Black Box Design

If rules are to help us at all (and generally, rules do not much help us to innovate) then probably the black box technique is likely to be a fruitful approach when a complex system is to be designed. We have already called this technique to our aid when defining the problem but it may help us further. Consider our problem in black box form as in Fig. 9.1.

Our problem is to devise hardware which will act on the stated inputs to produce our required output and it is to be assumed that, initially, we know of no hardware which will do this. We may, however, be able to devise useful sub-systems and components by working from the ends of our black box to the middle. We may be able to devise a system which with a given input will yield our required output and the original problem may be restated as Fig. 9.2.

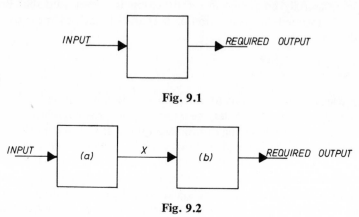

Fig. 9.1

Fig. 9.2

Since our problem of achieving the required output with X is already solved, our original problem has reduced to the design of a black box (a) which will act on our original input to produce X.

This technique is common in the design of control systems for consider the black box of Fig. 9.3.

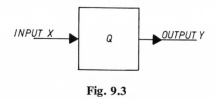

Fig. 9.3

The black box may be regarded as a device for generating the transfer function Q where

$$Y = QX$$

Those not familiar with transfer functions should realize that this expression does not mean X multiplied by Q. It is simply a shorthand way of saying that X is converted to Y by some process Q. Q may be a mathematical expression but it may also be a verbal description of a process.

We may find it convenient however, to create a system which consists of a large number of known elements (Fig. 9.4).

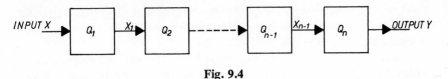

Fig. 9.4

so that

$$Y = QX$$
$$= Q_n X_{n-1}$$
$$= Q_n Q_{n-1} X_{n-2}$$
$$\overline{}$$
$$\overline{}$$
$$= Q_1 Q_2 - - - Q_n X$$

This technique is also used, although less consciously, by designers in many fields other than that of control systems.

An engineer asked to design an atomic power station may not, initially, know how to produce domestic electric power from radioactive ore. He will know, however, how to produce electric power from a steam turbine driven generator and he will know how to produce heat from a nuclear reactor so that his problem becomes the production of steam, using the heat produced in a nuclear reactor.

The pressure switch of Fig. 2.1 is, perhaps subconsciously, a product of this type of thinking. The designer knows that his bellows converts pressure into physical movements and his bought-out micro-switch converts physical movement into electrical switching. We must not press this example too far however because we must already be aware that Fig. 2.1 is unlikely to form the basis of a system that will meet the requirements of the specification we wrote in chapter 7.

Brainstorming

One technique for generating ideas, that is becoming fashionable, is that of 'brainstorming'. In this, a number of designers are made familiar with the problem and are then asked to contribute possible methods of solving it. The ideas put forward are listed without comment or criticism and the list may be as long as time or the number of ideas permit. Contributors are encouraged to make revolutionary or even wild suggestions so that there is a good chance of new ideas being put up; it is much easier to tame wild ideas than to turn conventional ideas into better ones. Further, the contributors should not be confined to designers with directly relevant experience, so that the chance of new ideas being put up is again enhanced. Contributors should also be encouraged to improve on the ideas of others. Research has suggested that 6–12 designers can produce about 50% more usable ideas by brainstorming than can a single individual (9a).

Perhaps the difficulty of brainstorming lies in the level of complication at which it can be applied. In value engineering, where a problem may have a comparatively simple answer, brainstorming techniques can be very profitable. Where the problem and its solution are both very complex, the results from a brainstorming session may be less useful but it is always worthwhile giving the method a try.

One difficulty is that a good brainstorming session will yield many possible solutions and ultimately all but one of these solutions must be rejected. Most of the solutions must be rejected without much analysis if we are to avoid spending too much money or time on the job so that brainstorming will be most effective in cases where many solutions are amenable to cheap, quick judgment.

There are variations on brainstorming. Synectics is a procedure studied by Gordon (9b) which involves a team of which only the leader knows the real problem. Some creative sessions are described by Gordon in which even deciding on a use for an invention is part of the function of the session.

Creative thinking is a personal thing and what works for one man may not work for another. A lot of research into creativity is going on and Matchett (9c) provides one example. Matchett has identified the following as typical catalysts in finding solutions to a problem:

  memories of past designs
  competitors' products
  deliberate doodling and day dreaming
  single words with rich associations
  self-questioning
  basic forms and archetypal symbols
  biological analogies
  science fiction
  irritation, anger
  complete quietness
  deliberate distortion of existing ideas
  technical reading
  trying to describe what one is attempting.
As mental directors, he lists
  private brainstorming
  use of formal proposals
  critical analysis techniques
  formal logic

scientific method
statement of objectives
definition of problem
definition of major obstacles
concepts of the structures of the design process
strategies proved successful in the past
systematic factorizing and charting.

Whether knowledge of this list helps us to create is doubtful but it does provide some indication of the difficulty of creating to order.

The best way for the student to study methods of creating solutions to problems is to try creating in a group, with and without certain aids and disciplines. A brainstorming session can be held with four or five participants. One man trying to think up new ideas will rarely think of as many as a group because a statement by one man will stimulate a train of thought in another. On the other hand, a dozen or more men creating out loud could simply lead to an ungovernable chaos. Incidentally, remember that a morning spent creating by six senior engineers will probably cost their employer well over £ 50 so that group sessions are not to be called up lightly in industry.

One of the greatest enemies of all design is the inability of people to get out of a rut in which they consider only the orthodox and commonplace. In the case of students, it is quite useful to loosen the mind on silly exercises such as:

'How many words starting with the letter W can you think of in 30 seconds?',

'How many uses can you think of in 5 minutes for a match?'.

Exercises like this have several advantages. They stimulate people into making wild suggestions and to the realization that wild suggestions are not always stupid; they also introduce helpful methods.

If in thinking of words beginning with W, someone suggests the word 'why', the suggestion is immediately followed by 'what', 'when', 'who' and then 'whether' so that it becomes apparent that ideas are worth investigating as classes. If suggestions for the use of a match peter out, they are likely to be stimulated again by looking for classes of use among the suggestions already made. If someone suggests using matchsticks for counting, are there any other educational uses? If someone suggests using matchsticks as toothpicks, are there any other uses to do with food? Cocktail sticks? Meat skewers?

Sometimes it is useful to set problems to show how people see constraints which do not really exist. A commonly used problem is that of drawing four straight lines through all nine points of Fig. 9.5, such that each line starts where the previous line finishes. This problem always seems difficult until one realizes that there is nothing to prevent a line extending beyond the square formed by the points.

Sessions may be tried with and without a chairman. A chairman may help by following leads which generate many ideas but he may limit the generation of ideas by imposing a discipline. It will almost always be useful to have a secretary and with more ambitious problems it will be desirable to have a tape recorder.

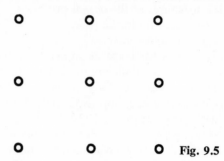

Fig. 9.5

When real design problems are presented it is sometimes productive to list classes of related problems. One could list similar problems that have been solved in nature or in other industries. One could list all possible sources of power, all possible means of propulsion.

Consider our pressure switch problem. A group of students could be asked to think of all possible ways of measuring pressure. Perhaps they would come up with a list which included the piezo-electric effect, a metal bellows, a metal diaphragm, a fabric diaphragm, a water (or mercury) column, etc. They could then be asked to think of all possible ways of making a circuit conducting or non-conducting. Many methods suggest themselves: mechanically operated contacts, Wheatstone bridge, transistors, thermionic valves, relays, etc. Methods of signal conversion could be considered: levers, bell cranks, fluid pressure, pneumatic or electronic amplifiers. If we continue to generate lists of possible solutions to parts of the problem we soon have a very large number of possible solutions to our main problem. Five methods of responding to pressure together with six methods of transmitting signals and five methods of making a circuit would give $5 \times 6 \times 5 = 150$ possible configurations to consider.

In fact, any worthwhile session would produce many ideas in each sub-problem so that we would very soon have thousands of combinations of ideas from which to choose a solution to our original problem. Of course, many of the combinations would be absurd: which is just as well, since we would not wish to analyse thousands of competitive proposals. One advantage of looking at sub-problems is that it removes many of the constraints caused by prior solutions of the main problem. This sort of session too can be run with only the chairman knowing the original problem so that other participants' minds are not constrained by existing methods of solution.

It is easy to see that a group brainstorming to solutions of our pressure switch problem would almost certainly consist of men who will be bemused by the picture of Fig. 2.1 if no action is taken to prevent this.

When all is said and done, however, heuristics is as big a problem to us as it was to the Greek mathematicians and we must all await inspiration before solving a problem. Years of experience help the problem solver, because the repertoire of prior systems that he acquires makes him less dependent on innovation, but sooner or later the designer is faced with a problem which requires innovation for its solution. The solution of such a problem is as likely to strike the designer in his bath as when he is seated in front of a drawing board. There is reason to believe too, that the most profitable innovations do strike the designer in a way that we cannot systematize. In almost any field, there are many men with experience, so that a problem solution which derives from experience is likely to be one which many designers could devise. New ideas which are capable of making a big killing in the market must have an original quality which cannot, by definition, be derived from experience.

It is possible that those qualities which make a man into a good, workmanlike designer, detract from his abilities as an innovator (9d). There is some evidence in fact, that innovation is more likely to result when the designer is not in the disciplined frame of mind with which he should confront analysis, but this should not worry us greatly because great innovation is much less often required than the workmanlike, detailed craftsmanship of design. Indeed unnecessary innovation can sometimes be the enemy of good design.

CHAPTER 10
Design Selection

When selecting the best design from those available, the ultimate criterion is profitability. Since we wish to reject unsuitable designs quickly and cheaply, we will first seek powerful reasons for rejection that will involve as little analysis as possible.

We may first apply the design specification (p. 33) as if it were a go/no-go gauge, thus:

Paragraph 1. Function. Will a brief consideration of the proposed system show that it cannot function?

Paragraph 2. Detailed Functional Requirements. Does a brief consideration of the proposed system show that the required performance cannot be achieved?

Paragraph 3. Operating Constraints. Does a brief consideration of the proposed system show that it cannot operate with the limited power available, that it cannot be worked by the operators available, that it cannot be maintained, that it cannot last the required life, etc.?

Paragraph 4. Manufacturing Constraints. Does a brief consideration of the proposed system show that it cannot be made with the available resources, that it cannot be proved with the available resources, that it cannot be made in time, that it cannot be made sufficiently cheaply, etc.?

Paragraph 5. Environment. Does a brief consideration of the proposed system show that it cannot stand the heat, the pressure, the vibration, etc. to which it will be subjected?

At this stage, we will not be looking for sophisticated reasons for rejecting a design; we will be attempting to eliminate proposed solutions which will not work because they break the laws of nature or which are likely to cost so much that we just cannot afford to go ahead with them. It is not usual for an engineer seriously to propose a problem solution which breaks natural laws although, from time to time, somebody suggests a problem solution involving perpetual motion or contravening the second law of thermodynamics. Occasionally, the engineer will have to solve a problem with a very advanced technical content and the feasibility of the solution may depend on our ability to distinguish very small signals, to differentiate signals from noise, to carry out the required number of operations in an acceptable time or on a number of other determinable, physical factors. In such a case, checking that the proposed solution does not break any natural laws may be a far from trivial task but it is unlikely, in the commercial world, that an engineer would work in a highly specialized, scientific field without being aware of its special problems.

Most engineers, dealing with most of the problems that they are asked to solve, will not generate solutions which are physically impossible to realize, for the history of engineering teaches us that practically any proposed solution to a problem can be made to work with enough expenditure of time and money, so that our criterion for rejecting proposed solutions to our problem is likely to be a commercial one. It would be comforting to believe that designers actually go through a deliberate process of sifting to eliminate proposals which will not work and of comparison to ensure selection of the cheapest of the survivors. In fact, the more one analyses a design, the more one finds sub-problems to solve and each solution to a problem seems to bring fresh problems in train. Even when the designer is convinced that several designs will meet the performance requirement, who is to say which is the cheapest? Who is to say whether a nuclear power station is better than a coal-fired or oil-fired power station? Even the prices per unit of electricity cannot be compared until the write-off lives of the power stations are determined. With major systems there must always be an arbitrary element in the judgment because of the complexity of the calculation, the accuracy with which probabilistic events can be forecast and the time available to make decisions.

Luckily there are some simple but powerful criteria which can be applied at an early stage and we have already identified:

does not obey the laws of nature,

does not meet tolerances,

will not work 'off-design',

will not withstand environment

and, in fact, the last three of these criteria have already been demonstrated by our example of a pressure switch design.

Most of these criteria require the ordinary processes of analysis familiar to engineers but there is seldom time to evolve definite numerical answers which enable an unerring choice to be made. Marples (10a) has shown that designers, consciously or unconsciously, use a decision tree approach to the comparison of alternative solutions.

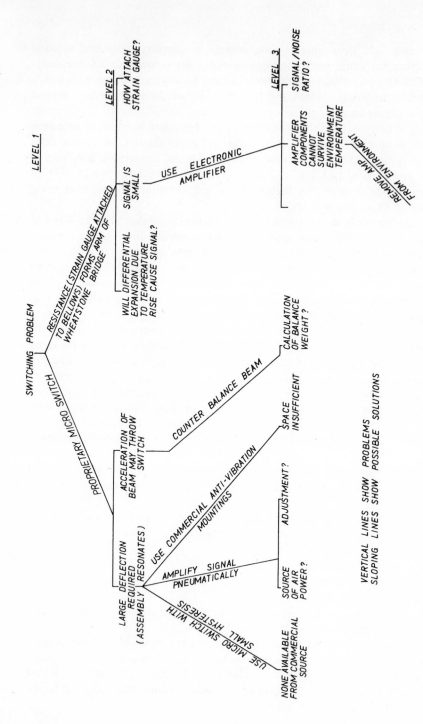

Fig. 10.1 Part of decision tree of part of pressure switch problem

Consider, again, our pressure switch. Let us assume that company policy (available skills, background knowledge, etc.) has suggested the use of a stainless steel bellows to convert the pressure signal into movement. We now wish to use this deflection to break and make an electrical circuit. Using the Marples convention we show part of a reasonable decision process on Fig. 10.1.

In this decision tree we see that one or more solutions have been proposed for each problem and that each proposed solution, in turn, leads to further sub-problems. Further, each proposed solution to a problem requires analysis and perhaps experiment to show whether it is feasible and what new problems evolve from its use.

We see that at level 3 we are able to eliminate certain proposals. We cannot use a smaller micro-switch than the one already considered because we cannot find one in the makers' catalogues. We cannot mount the pressure switch on commercially available anti-vibration mountings because the resulting system would be too large.

Even when we have eliminated the unacceptable we still have many possible solutions to our problem which must be compared. The designer must therefore attempt to judge the cost of the possible alternatives. Remembering that we are restricted to about four man weeks to prepare a brochure for submission to the customer, we realize that the designer can hardly expect to do more than rank the possible solutions in order of merit. The most that the designer can afford to do is 'order of magnitude' calculations. We know that no bellows will have a much larger effective area than about 1 sq in or it will break the overall size constraint. We know also that with any feasible moving mass the stiffness of the assembly must exceed about 1250 lbf/in if resonance is not to occur below 500 c.p.s. Since we have a tolerance band of $\frac{1}{2}$ p.s.i. in our measured pressure, we know that a movement of well under 0·0005 in will be all that we can use.

These order-of-magnitude figures enable us to consider the proposed solutions to problems at level 2 of our decision tree.

Among the proposed solutions are:

> *a pneumatic amplifier,*
> > *which would have an input signal of 0·0005 in deflection and required outputs of at least 0·006 in deflection and 6 oz load.*

> *an electronic amplifier,*
> > *in this case a signal has been provided by a strain gauge stretched by 0·0005 in. The output must be sufficient to operate a relay or solid-state switching system.*

There are, of course, more alternatives, but only the above two are shown on the decision tree. Of these two alternatives, the firm concerned has some experience of pneumatic amplifiers but none of welded strain gauges so that the pneumatic solution would probably be ranked above the electronic amplifier, in order of preference.

It is to be noticed that any judgment has been made from the point of view of physical feasibility (almost, in fact, out of desperation) although function, operating and manufacturing constraints and environment have been considered in making a decision. This example illustrates one of the designer's dilemmas. Often he will choose what will eventually be seen as the wrong solution because of the time, money and chance involved, but his immediate task is to satisfy a customer. Many of the great breakthroughs in engineering were achieved by men who lost money on the venture or failed to satisfy the immediate customer. Eli Whitney is credited with the invention of mass-production methods but few historians point out that he took so long to bring the idea of interchangeable parts to a workable state that he failed to deliver the muskets ordered by the U.S. Government.

In very trivial cases it may be possible to calculate the pay-off of each policy. This seldom occurs in commercial problems but will be discussed as a special case of optimization in a later chapter.

Although we have used our design specification as a check list, we have concerned ourselves mainly with the physically feasible. A more detailed reading of the specification tells us that we must also consider, at a very early stage, other problems of life, cost, operation and manufacture. Among these problems are:

Safe Failure

Consideration of life and costs leads us to think of the possibility that there may be certain types of failure, the results of which will be catastrophic. This could be regarded as constituting a very high cost so that the possibility of disastrous failure may be sufficient reason for rejecting a proposed design. Failure analysis is, in fact, very powerful because it can be applied at a very early stage of design. Since the techniques of FAILURE (and RELIABILITY) analysis require more than a passing mention, they are discussed at greater length in chapter 16.

Time

A system may have no real value if it cannot be delivered to the customer on time. Usually a system will decrease in value the later it is delivered because obsolescence is as much a function of competition as of wear-out. Several types of propeller-driven aeroplanes had their useful revenue-earning lives very much reduced, because late delivery meant that they were competing too soon with the next generation of pure-jet aeroplanes.

Sometimes, however, there will be an absolute date, beyond which the system is not acceptable to the customer. An exhibition stand is of no value if it is delivered to the customer after the exhibition for which it was intended.

Our design specification for the pressure switch specified the delivery date for the prototype systems and also the minimum development standard at which they could be accepted. We have already discovered that vibration will be a major difficulty in this design and so we will be committed to solving this problem and demonstrating in the laboratory that the system will survive 100 hours of vibration before delivery in three months time. If we allow for the probability that mistakes will be made and there will be more than one attempt to survive 100 hours and if we also allow for the low confidence level permitted by testing only one or two systems, we realize that the time required by the tests alone presents great difficulties in meeting the due date.

Sometimes the customer will write into the contract great penalties for a manufacturer who does not deliver by an agreed date.

The designer must, then, demonstrate (to himself and to his customer) that it is reasonable to expect delivery of the system by an agreed date and such techniques as CRITICAL PATH METHODS (see chapter 13) must be used to demonstrate the feasibility of delivery promises. Later, similar techniques may be used in the management of the manufacture of the system.

Cost

Prominent among costs are design and manufacturing labour. The cost of raw materials is usually not difficult to calculate once a scheme has been prepared but, particularly in highly technical industries, the labour content of a system is far more significant than the raw material content. The designer must then demonstrate to himself and to the customer that the labour content of his system is within an acceptable cost. The assessment and control of labour is related to time analysis by Critical Path Methods and is discussed in chapters 14 and 15.

The costs of maintenance must also figure in any judgment between competing systems. Many systems have failed to be commercially acceptable because either they needed too much maintenance or they were so designed that, when needed, maintenance was either impossible or too expensive. The problem of time between failures is one of RELIABILITY, discussed in chapter 16, while the problems of the MAN/MACHINE INTERFACE are discussed in chapter 20.

Optimization

When the proposed systems have been checked by coarse judgments, those that survive must be compared by finer analysis. At some stage we will either be left with only one acceptable system, which will be proposed, or, where several systems survive and the cost of judging between them is too great, an arbitrary choice is made of a solution which will be proposed.

Mostly our choice of system lies between several which are distinctly different and we must calculate the profit from each candidate to a point at which we are prepared to select one system rather than another. This was the case with the pressure switch but, as a further example, we may consider a steam engine, a petrol engine or an electric motor as competing systems to power a motor car. In order to select one system from the three offered, we would have to analyse each until relative costs became apparent. Then a choice could be made.

In some situations, however, competing solutions to our design problem are numerical variations on a single theme. A framework designed to support a loan may consist of struts and ties and differ only in geometry from an infinity of frameworks designed for the same job. In such a case we may call upon the tools of optimization, HILL CLIMBING, LINEAR and NON-LINEAR PROGRAMMING (see chapters 17, 18 and 19). In some situations, the choice is not a simple one. If we were designing a power station, it may be that in a given situation one competing design is better than another. It may be, however, that from a national point of view, several power stations will be required and that the optimum solution will be to accept several different power stations of several different types. Where the optimum solution is a product 'mix' of systems rather than a single system, the tools of optimization will be required to determine the best mixed strategy to adopt.

The system will normally be proposed in a manner calculated to demonstrate to the customer that it is a good one. What the customer needs to be told will vary from case to case but in this and chapter 8 we have seen that any brochure must contain:

a demonstration that the system will work: in design conditions, in necessary 'off-design' conditions, within required tolerance, in the environment,

a demonstration that the system can be made economically and on time,

a demonstration that the system can be operated economically,

a demonstration that the system can be maintained economically,

a demonstration that the required safety, reliability and life will be met in the environment.

The ways of demonstrating the merits of a proposed system may vary from real-life demonstrations of hardware to mathematical dissertations. Brochures may be typescript leaflets or seven-volume works of scholarship. In almost all cases however, the objective is to convince the customer with reasoned argument that he will profit by buying the proposed system and in almost all cases it is necessary to do so in the face of competition.

The tools that the designer will use will include the classical physics normally taught to engineers, but in addition he will be expected to have an acquaintance with Failure and Reliability Analysis, Critical Path Methods, Resource Allocation, Accountancy, Modelling, Optimization, Ergonomics and Work Study.

Exercises and Subjects for Discussion on Part 2

Exercises 1, 2 and 3 should be solved by discussion between students in groups. An afternoon could be devoted to such a group exercise.

1. The eye and the camera have much in common although one occurs in nature and one is man-made. The ball and socket joint is another mechanism which occurs both in nature and in man-made systems.

Think of some other systems occurring in nature which either have been, or could be, copied by man.

2. What has a spring/mass system in common with an L-R-C electrical network? Devise

(i) a mechanical system:
(ii) an electrical system:
to add two quantities
to multiply two quantities
to integrate a function w.r.t. time
to differentiate a function w.r.t. time.

3. Put together some black boxes into a system which will show the average speed of a vehicle, calculated from the commencement of a journey.

Given the speed of the vehicle, every minute for a ninety minute journey, how would you determine the average speed during the period between the start of the journey and any given time before the end of the journey?

Suggest both mechanical and electrical sub-systems which could be used.

4. Get brainstorming sessions going with groups of about six students.

How many uses can you think of, for a housebrick?

How many ways can you think of, for getting a man from the top of the Post Office Tower in London to the top of the Empire State Building in New York?

How many ways can you think of, for centrally heating a house?

Compare the results achieved by groups working with and without a chairman and with and without methodical procedures.

The following exercises are for individual students or for small project groups. Any one of the exercises may take several hours a week for two or three weeks.

5. Write a design specification for heating the houses of a new town. Outline first, the significant features of the town that you have chosen.

Suggest means of meeting the specification that you have written.

Write a short brochure, or leaflet, proposing your system to the builders of the new town.

6. Write a design specification for a town car. State the town for which you are specifying the car and outline its significant features.

Suggest means of meeting the specification that you have written.

Try thinking of different sources of energy, different types of prime mover, different methods of propulsion, different types of road, different methods of steering as separate problems. What combinations of these solutions form good solutions to the original problem?

Write a newspaper advertisement or a 1000 word newspaper article, designed to sell the car you propose.

7. Suggest means of meeting the specification that you have written in answer to question 5, page 49.

Draw a decision tree to show the arguments that led you to select a particular solution to the problem.

Write a brochure intended to sell your design to the Chief Constable.

8. Suggest means of meeting the specification that you have written in answer to question 8, page 52.

Draw a matrix of three of the factors affecting the problem (possible power supplies, possible cleaning methods, possible uses, possible methods of locomotion and any other factors you think may be relevant). It is useful if the problem can be discussed by a group of students who do not know the basic problem (i.e. the discussion will be led by a chairman who is the only member of the group who knows that they are trying to invent a cleaner). If, in another group, the students know the problem but use various puzzles and opening problems to open their minds, the results of different methods may be compared. Either score the entries in the matrix or rank the entries in order of excellence.

Draw a decision tree to show the arguments that led you to select a particular solution to the problem.

9. Suggest means of meeting the pressure switch specification on p. 36.

Write a brochure designed to sell your proposal to the customer.

Bibliography to Part 2

9(a) American Society of Tool and Manufacturing Engineers, *Value Engineering in Manufacture*, Prentice Hall, Englewood Cliffs N. J., 1967.
9(b) W. Gordon, *Synectics*, Harper & Row, New York, 1961.
9(c) E. Matchett, Chartered Mechanical Engineer, **15**, 163, April 1968.
9(d) D. Mackinnon, 'The Creativity of Architects', contribution to *Widening Horizons in Creativity*, edited by C. Taylor, Wiley, 1962.
10(a) D. Marples, *Decision of Engineering Design*, Institution of Engineering Designers, London, 1960.

Further Reading

A. Osborn, *Applied Imagination*, Charles Scribner's Sons, New York, 1957.
J. Alger and C. Hays, *Creative Synthesis in Design*, Prentice Hall, Englewood Cliffs N.J., 1964.
S. Gregory (Ed), *The Design Method*, Butterworths, London, 1966.
M. Starr, *Product Design and Decision Theory*, Prentice Hall, Englewood Cliffs N.J., 1963.

Part 3

MAKING THE HARDWARE

Engineering design requires many skills and much knowledge and no ordinary man can expect to have all the abilities necessary to produce a good design in any but the most trivial situation. A design team will include many specialists and good design is as much a problem of ensuring co-operation and communication between small groups of specialists as it is a technical problem. In industry, it is a matter of everyday observation that the members of a design team do not understand the working of the whole system. The performance engineer will not even know that the jig and tool designer is part of the same organization, the prototype shop fitter does not realize that his main function is to supply information to the designer. In the university, the undergraduate rarely understands the context in which he will be expected to work.

There is little written about the structure of a design team. Burnham and Bramley, 1957, have a useful chapter on the Control of Production, which describes communication in some detail. Hicks, 1966, has some comments on communication also. The best descriptions of the problem are probably found in the Drawing Office Manuals of the larger companies.

It is important that students be taught the reasons for formal documentation and they should be discouraged from using sketches or verbal information as means of communicating with workshops.

CHAPTER 11
The Team

We have defined the designer's job as supplying the manufacturer with the instructions necessary to build the system and that job is not done until the system which results from those instructions really will meet the specification. Common experience tells us that the designer has not finished a job when he has merely invented an apparently good solution to the customer's problem. The system has to be built and shown to be a good system.

Usually, when manufacturing instructions are first issued it is found that they are wrong and the manufacturer is unable to build the system without modifications to the instructions.

When the instructions have been corrected and the system built, it is usually found that the system will not work. Once again, the system must be modified until it does work.

Commonsense tells us that if we are to expect trouble with the first systems to be built, it may be economical to build a small number of prototypes, with which to solve as many manufacturing and development problems as possible, before embarking on the production of a series of systems for the customer (this does not, of course, apply where the total number of systems to be produced is only one or two—as, for example, in the case of a nuclear power station or a large ship). Again, if we are expecting to experiment with manufacturing methods, it is reasonable for the prototypes to be built in a different shop and by a different class of labour from that of the series production systems.

When a system has been built and we wish to check that it functions satisfactorily, we may allow it to go into service, see how well it functions in real life and modify it if it fails to perform satisfactorily. Clearly this is not always a desirable thing to do for it may lose friends among customers and could be too expensive a procedure for either the customer or the manufacturer to contemplate. Situations differ; an aeroplane must be checked on the ground, in test rigs and in the air by test pilots, etc. so that the manufacturer is as sure as he can be of its safety before passing it on to the customer; in the case of the motor car however, it may be permissible for much of the development work and rectification of design errors to take place after sale, using the customer as a test engineer.

Normally, the first systems to be produced will be tested, as far as possible, in the laboratory, to find as many design errors as possible. These design errors will be modified and the modifications proved before any systems are put into service.

In addition to the basic, creative design office then, we may have A PROTOTYPE SHOP, A PRODUCTION SHOP and A DEVELOPMENT DEPARTMENT. These manufacturing and testing departments must be told what to do since they have no knowledge of the designer's intentions until they are told. Usually instructions to the Prototype Shop consist mainly (although not exclusively) of drawings which specify the geometry of each detail part and of the assemblies. It is assumed that the Prototype Shop contains a high proportion of skilled craftsmen who will work directly from drawings, who may even design and build any necessary tools, and who will generally make parts by largely manually controlled processes. The drawings and other manufacturing instructions must be produced, however, from the scheme proposed by the designer and generally a DRAWING OFFICE exists for the purpose of converting the designer's original scheme into detailed manufacturing instructions.

The instructions to the Development Department will normally consist of a schedule of tests, devised by the designer to assure himself that the system operates satisfactorily. Naturally, as time goes on and tests reveal faults the development engineers will reasonably suggest tests, but initially only the designer is able to suggest and issue a schedule of useful tests.

Eventually a system will be built in a Production Shop and probably a batch of systems will be built together or a continuous line of systems will be produced. Generally the workers in a Production Shop are less skilled craftsmen than those in a Prototype Shop and their instructions may not take the form of drawings requiring skill to interpret. Further, batch or flow production may involve special techniques, requiring special knowledge which it is assumed that the designer does not possess. Again, the manufacture of a number of components by men of limited skills will justify the design of special tools to simplify and cheapen production. The drawings that have already been issued to specify the geometry of each part and the other manufacturing instructions may therefore be translated into simpler instructions defining simple manufacturing processes and the tools and equipment needed to perform them. A department making this translation may be called a PLANNING DEPARTMENT. (Although many names such as Production Control, Methods, Jig and Tool may be used in describing such a department).

Throughout all the processes of manufacture and testing, it will be necessary to check that instructions have been properly executed, that components are in fact as required by the drawings, that test requirements are properly met, etc. Such tasks of inspection may require special skills and special equip-

ment and may well be the function of a QUALITY CONTROL DEPART-
MENT (again, other terms such as Inspection Department may commonly
be used).

However well the system has been tested, experience in service will show
faults and failures that can or must be eliminated by modifications to the
design. SERVICE ENGINEERS must be available to follow the progress of
systems in service so that faults which are shown by service conditions may
be rectified by modifications to the design.

Only after service has demonstrated that the system performs as required
of it and manufacturing instructions may be guaranteed to reproduce such
a system is the designer's job done. The organization involved in reaching
this objective may be quite complex, may contain many special skills and
much specialized knowledge. It many cases the designer will have less knowledge of a particular field than his colleagues but it will be his function to
co-ordinate all the diverse skills so that his objective is achieved at
minimum cost.

CHAPTER 12
Communications

We have identified, in the design process, the following actors: Customer, Designer, Drawing Office, Prototype Shop, Development Department, Planning Department, Production Shop, Service Engineers, Quality Control, and it is clearly essential that these separate departments should communicate efficiently with one another. One of the costs of design is the cost of this communication. Even greater cost is caused by any failure of communication.

It is the designer's job to ensure that information is transmitted effectively, quickly and cheaply and he cannot do this if the organization is badly managed.

If we sketch the principal relationship between these departments we obtain Fig. 12.1 and in this diagram we have:

The Design Specification (document 1)

This document carries information about the problem to the designer. We have already discussed this information and have shown that the required information is best transmitted in a formal way via a design specification, and we have also seen that the required information is rarely made directly available by the customer in his first communication with the designer. Usually the customer makes known his needs informally and inadequately, then the designer produces a design specification which is submitted to the customer for approval and any necessary modification. By considering the customer's comments, the designer is able to produce a more precise design specification which, in turn, is submitted to the customer and the whole process is repeated until both customer and designer agree that the specification really defines the problem. Before this stage is reached, information may circulate from customer to designer and designer to customer many times. Each time the design specification is modified and submitted to the customer, its issue number is raised until, finally, at an agreed stage, the design specification is sealed, with an identifying number (common to all issues) and an identifying issue number. This identifiable specification will form the basis of any financial agreement between the customer and the designer (or manufacturer).

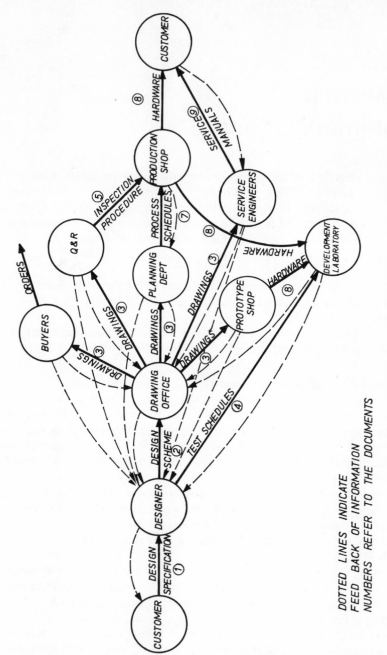

DOTTED LINES INDICATE
FEED BACK OF INFORMATION
NUMBERS REFER TO THE DOCUMENTS

Fig. 12.1 Information flow in a design organization

Important points to notice are:

the formal manner in which information is transmitted and

the fact that feedback is as important as the forward flow of information.

An example of a design specification form has been given in Fig. 7.1.

The Design Scheme (document 2)

The designer, as has been described in earlier chapters, must invent a system which will meet the requirements of the design specification. The system must be described in sufficient detail to instruct the draughtsman but before this information is transmitted to the draughtsman, the customer must have agreed that he likes the system enough to buy it. We are therefore involved, once again, in a flow of information from designer to customer and from customer to designer (as the customer requires modifications) and eventually, if the designer's efforts are successful, the customer places an order. Usually this loop is less formally controlled since the design specification can carry all the formality that is required. If the designer's system proposal contains elements which the customer wishes to be regarded as mandatory then these mandatory requirements can be embodied in the design specification.

The transmission of the information to the drawing office must, however, be formal. Much of what has been supplied to the customer will be useful matter to supply to the draughtsman but much more detailed information will also be required.

The designer is responsible for the way in which the system is manufactured and for the cost of this manufacture. The draughtsman is not responsible for the geometry or material but for translating the information supplied by the designer in a way that may be easily understood by a prototype shop operative or a planner. Normally the information is supplied by the designer in the form of annotated drawings. From the drawings, the geometry should be clear to the draughtsman, while from the notes, materials and any special comments on manufacturing methods should be apparent. These annotated drawings, sometimes referred to as design schemes, must be formal, recorded and identifiable. On these schemes depends much of the reputation (and hence earning power) of the designer. The information produced by the designer does not consist of ideas and suggestions but of precise instructions and in principle, at least, those instructions should contain all that is necessary to manufacture the system. Fig. 12.2 is a part of a typical design scheme.

Some designs are clearly better in conception than others in the sense that they must result in lighter, better performing, easier handling or in other ways cheaper systems but at the stage where detailed manufacturing instructions are being formulated, the broad choice of principle has been made. At the stage we are now considering, the system can only be made

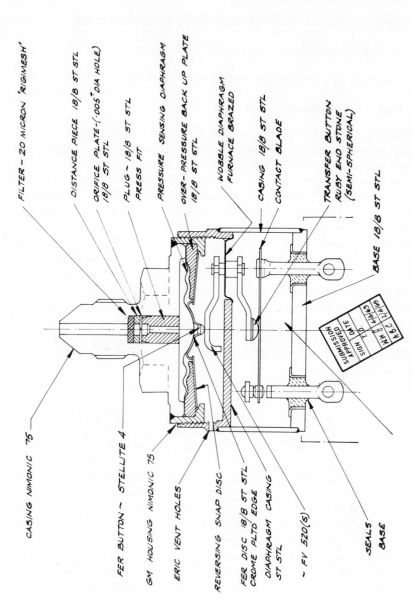

Fig. 12.2 Fragment of a design scheme

worse and it can be made worse by faults of detail; such elementary faults as failure to lock threads adequately, failure to ensure that electrical connections are firm, failure to consider the best sequence of manufacturing operations, failure to consider adequately the manufacturing tolerances etc. A draughtsman has a certain skill and certain experience, and in many cases the designer may leave decisions to him, but ultimately the designer is responsible for the cheapness of manufacture and the cheapness of operation of the system. The designer must decide (and accept responsibility for) which decisions may be left to the draughtsman and which he must take himself. Those decisions that he takes himself must be clearly transmitted and recorded. The design scheme must therefore carry an identifying drawing number (and issue numbers, as changes are made). Because designer and draughtsman sometimes work in the same room and are in daily, conversational contact, the formal demonstration of the designer's instructions is often omitted. This is dangerous, not only because informal instructions may be misunderstood but because it permits the designer to avoid the discipline imposed by formulating and studying his own proposals.

Few designers can be expected to have as much skill in every subject as the specialist. He will not, for example, be as knowledgeable about machining operations as the foreman of the machine shop; he will not know as much about inspection techniques as a senior quality control engineer; he may know less about the capabilities of a welder than the foreman of the prototype shop and so on. There is no virtue in the designer's taking decisions without the advice of those who have specialist knowledge and a commonly adopted procedure is for the design scheme to be submitted to the relevant specialists, for their comments, before it is passed on to the drawing office. This may be done by a formal procedure whereby each specialist is sent a copy of the scheme for examination and comment (note the panel for signature on the design scheme of Fig. 12.2) or it may be done by calling a meeting of those concerned, to discuss the scheme. The decisions are still to be made by the designer. The opinion of one specialist must be weighed against the opinion of another. Weight must be given to those opinions but they will probably sometimes conflict and sometimes reveal too narrow a viewpoint. It would be reasonable for example, for a production engineer to request that for cheapness a tube be made by rolling and welding sheet metal while the quality control man might require that the same tube be drawn in order to avoid the expense of inspecting the weld. Both men have a case but the designer must weigh the costs of the two possible methods and must then make the appropriate decision.

We see then that we have further complex loops in which information circulates. Between the designer and his colleagues who have useful specialist knowledge, the schemes will circulate until the designer believes that all criticisms have been met and all useful advice taken (this procedure may resemble the work of a Value Engineering committee—see Appendix).

When this stage is reached, the scheme will be given to the draughtsman so that detailing may commence. No doubt, as the draughtsman works he too will find points to criticize or recommendations to make. Such comments should be made to the designer so that the scheme may be further modified if the designer thinks it desirable. If changes are made without bringing the scheme up to date, it will be difficult to consider any analysis, development or manufacturing problem since only the scheme will summarize the complete requirements.

One area in which both the designer and draughtsman may offer cost saving is in the use of standard parts. Rarely does a design break new ground in every part. Usually, much of the design will be derived from experience and any possible saving to be derived from previous experience must be exploited. Benefit comes from using standard parts (or processes) when those parts have already been drawn: saving the cost of drawing; have already been manufactured: saving the cost of finding and solving manufacturing difficulties; have already been developed: saving the cost of further development; and have already been proved in the field: saving the cost of failure in service. Fig. 12.3 shows a page from a typical catalogue of standard parts.

Unfortunately, in some very small local process associated with design, it may be easier to produce a new idea than to look for a better one that has already been proved. A draughtsman could, for example, draw a new 'O' ring or a special screw with less effort than would be required to search a catalogue for an existing, proved, suitable part. Various proprietary systems are available for increasing the use of standard parts but any system must eventually depend on the accurate cataloguing of those classes of component which can usefully form standards. It is also necessary to ensure by good management and organization that standard parts and processes are used wherever that use would result in cost saving.

Drawings (document 3)

The drawings (and other documents) which define the geometry of the system and any special processes in its manufacture are, in turn, instructions to other men who have work to do. To the draughtsman and to the designer, the most obvious function of the drawings is to supply information to the operative on the Prototype Shop Floor. Clearly, the Prototype Shop Foreman is responsible for the efficient operation of his shop and he will decide which operatives will do which jobs, which machines will be used, how long each task should take, whether an outside supplier should be asked to make some of the parts, where he can best obtain his raw materials and many other decisions necessary to make the Shop profitable. Having made those decisions, however, it should be possible to give the drawing to the operative and that drawing should carry all the instructions he requires to make the part (or perform some of the operations required in manufacturing that part). Frequently, because of faults in the design, this is not the case, but the design

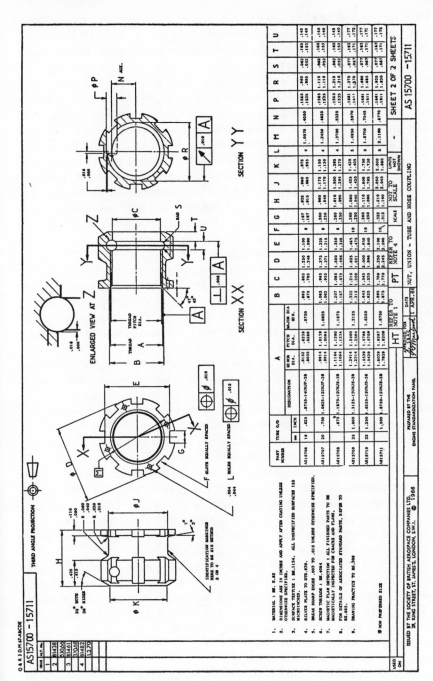

Fig. 12.3 Typical catalogue of standard parts

must be rectified by rectifying the information. It sometimes happens that a craftsman, proud of his skill, makes something in spite of the drawings but this can lead to the perpetuation of false information unless the knowledge of the error is fed back to the designer.

The same information must also be transmitted to the Planners. More correctly, the Prototype Shop seek out errors in the information, which is corrected and the corrected information acts as instructions to the Planners. The drawings have already been shown to be adequate for the Prototype Shop operatives so that one would expect them to be adequate for the Planners, whose task is to design the cheapest method of producing the number of systems required by the customer. Extra information is, however, required by the Planner. He must be told the number of systems required. Building a million motor cars would require quite different methods from building 10 motor cars. If the Sales Department is efficient it will tell the Planners, not only the immediate requirement, ordered by the customer, but also the prediction of future demand.

The drawings must also be transmitted to the Quality and Reliability Department. When it comes to manufacture, whether in the Prototype Shop or in a Production Department, inspection will be necessary. Much inspection is straightforward; an Inspector armed with adequate drawings will be able to check whether a component has been made to the required dimensions, within permitted tolerances, by simple measurement of the component, but even such simple inspection may benefit from planning. The time taken to inspect a component costs money and it may well be worth while to make available simple inspection gauges which will speed inspection. A diameter which is to be 1 in ± 0.002 could be measured on a micrometer but a simple go/no-go gauge would be much quicker to use.

In any case, not all inspection is simple and the inspection process may require planning by an expert. For example, special X-ray techniques may be required, special electronic-testing consoles may have to be designed, special sampling techniques may reduce the number of parts to be inspected, special government or customer standards may have to be maintained, expensive special equipment may have to be purchased. Such functions are the duties of Q and R Specialists who can only do their work if they have adequate information in the form of detail drawings, etc.

Sometimes the designer will have made decisions without adequate knowledge. This could be quite permissible where a surface finish or a tolerance has to be specified even though there is not sufficient knowledge available to permit the designer to forecast the results of his decision. In such a case, the Q and R man could construct an experiment to be carried out on a variety of tolerances so that, at least early in production, the significance of the designer's decision will be understood.

Both Planners and Q and R men may find that they wish to modify the design in order to make manufacture or inspection cheaper. This should

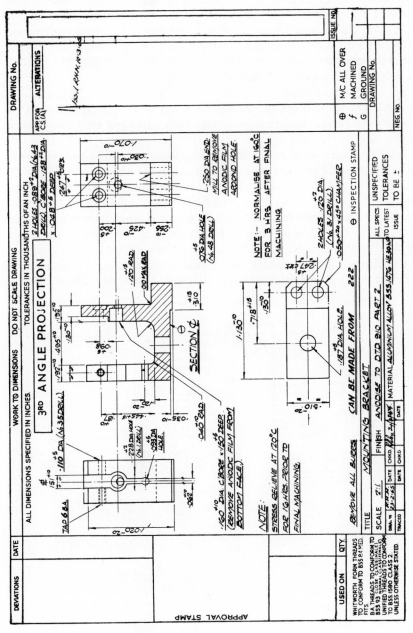

Fig. 12.4 Typical detail drawing

only be done by proper feed-back of comments to the designer so that, if modifications are agreed to be desirable, all the relevant paperwork will give correct instructions.

In many companies, the buying of raw materials is done by a special team of buyers with good knowledge of supplies and of the market. The buyer's job is to buy what is necessary at the best price consistent with delivery and in order to do this he too needs the drawings.

Someone in the company must decide how the system is to be maintained or repaired and what the spares-holding policy of the company should be. Usually the Service Engineers undertake these tasks and their information is provided by the drawings.

Usually, the drawings required to make a system are many and it is useful to have a list of all such drawings. Generally, anyone who needs the drawings will be provided with a Schedule. Also required will be a Manufacturing Instruction so that all the departments concerned will know that an order has been received, how many systems are required, what drawings are necessary, any permitted departures from drawing and any other information thought desirable. Figs. 12.4, 12.5 and 12.6 show a typical detail drawing, a page from a typical schedule and a typical manufacturing instruction.

When things go wrong, however, as they will, it is the duty of the Q and R

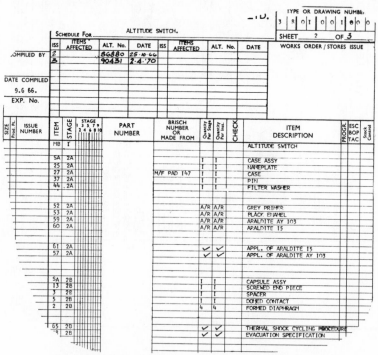

Fig. 12.5 Page from a typical schedule of drawings

man, the Buyer, the Planner and the Service Engineer to feed back information to the Designer and to the Draughtsman, so that corrections may be made and the drawings modified to show the correct information.

FROM DRAWING OFFICE –
TO :- MR.
COPY TO :-

DATE :- 29·2·68
SHEET Nº :- 6
Nº OF SHTS :- 6

TYPE			SUBJECT 4" BUTTERFLY VALVE.	
REQD. FOR	BENCH TESTS	FLIGHT		JOB Nº

ACTION REQUIRED	SCHEDULE OR DRAWING Nº	Nº OFF	DATE REQD.	REASON FOR INSTRUCTION
ISSUE 14 MANUFACTURE HOT RIG ADAPTORS. TO	DRG PTR 1372. ISSUE 1	1 SET	4-3-68	ACTIVITY CODE 59.
ISSUE 15 MANUFACTURE SEALING RING PART Nº 5106002228 MAT:- HAYNES 25. (SHEET) COLD ROLLED CONDITION	ISSUE 1.	2.	15-11-68	FOR TRIAL TEST PURPOSES.
ISSUE 16 MANUFACTURE AND ASSEMBLE. VALVE SER Nº 12592 DJ TO CARBON BUSHES TO BE MADE TO MATERIAL EY.9.	5279. ISSUE 1.	1	22-11-68	FOR DEV PURPOSES.
ISS. 17. 24. 3. 70 MODIFY (a) SEALING RING PART NO 5106.002.221 TO (b) ASSEMBLE TO VALVE S/N 12594 DJ. TO	SK 101040. Drg 5106.002112	1.	3/4/70.	PASS TO DEV FOR TORQUE TEST.
ISS 18 5-5-70 STRIP & VISUALLY EXAMIN BUTTERFLY VALVE TYPE 5106002000 SER NO 12594 DJ.		1		CONTACT DEV LAB.
ISS 19. 3. 6. 70 MODIFY. VALVE S/N 12596 DJ. TO	SK. 6090. ISS.1.	1.	16. 6. 70.	

PRODUCT GROUP Nº :- 1

ISSUE Nº	14	15	16	17	18	19		COMPILED
DATE	29-2-68	5-11-68	14-11-68	24.3.70	5.5.70	3.6.70.		
ISSUE Nº								APPROVED
DATE								

Fig. 12.6 Typical manufacturing instruction

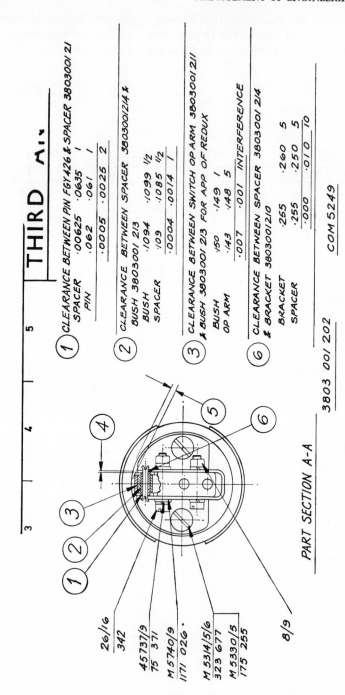

Fig. 12.7 Fragment of a tolerance study

Development Test Instructions (document 4)

Only the designer knows the reasons for the geometry and materials that he has specified in his scheme. Unless the original design problem was trivial, it is unlikely that the system, when built, will work as required. The system will have to be tested but the tests must do more than say whether the system works or not, they must diagnose any trouble and help to suggest a cure. The best procedure is for the designer to consider the required properties of each component and design tests which ensure that each component possesses these necessary properties. Often this is simply done by inspection of the part but often tests are required in laboratory conditions. Proof that a surface finish on a joint will give a sufficient seal, proof that a moving part has a sufficient fatigue life, proof that a required level of electrical insulation will be met, even in humid conditions, proof that a finish will give the required protection against corrosion in a hostile environment, are typical of problems which require skilled laboratory testing. Such problems should have been apparent from the design specification so that the designer should have been aware of them, have attempted to solve them and have called up the tests appropriate to proving his success. At the scheme stage and again at the detail drawing stage, the designer will be concerned to build, in imagination, the system from its components. He will carry out tolerance studies to check that parts mate, he will check that dimensions can in fact be met and measured, he will check that details can be assembled into sub-assemblies and assemblies. He will, in short, conduct a formal exercise to ensure that what he has designed can be built and inspected. (Fig. 12.7 is a fragment of one such tolerance study.) He will then submit, in his imagination, the system and its parts to all the conditions that it will meet in service and, by calculation, or otherwise, convince himself that what he has designed will perform as required.

From these two studies will evolve the test programme, which he will supply to the Development Department.

Once again, as tests are carried out, failures will be detected and the Development Engineers will gain knowledge of the system. Faults and the further planning of experiments must be fed back so that experimentation is done as required by modified schedules and, at all times, information will be found to be up-to-date and correct, as far as knowledge permits. Fig. 12.8 shows a page from a typical Development Test Schedule.

Production Test Instructions (document 5)

When the system is being made, either as a prototype or in the production shop, the details, sub-assemblies and assemblies have to be inspected. Nor-

Test Schedule Serial No. 1.
Issue No.1

-5-

5. Vibration Tests.

5.1. The test valve will be mounted on the vibrator by its normal attachment
points. Any additional stays or straps must be avoided and any additional
connections to the unit should impose the minimum restraint.

5.2. The test valve will be subjected to a resonance search in each of three
mutually perpendicular directions. Resonance of the test valve will be
determined by varying the frequency of applied vibration slowly through
the specified range at vibratory accelerations as shown in THG.1614.

The resonance search will be carried out:

(a) With an inlet pressure of 200 p.s.i.g and a pressure of 2.1 p.s.i.g
applied to the tank sensing line.

(b) With an inlet pressure of 20 p.s.i.g and a pressure of $1^{\pm}$ 0.1 p.s.i.g
applied to the tank sensing line. The flow from the valve will be
restricted to approximately 12 s.c.f.m.

If resonance occurs the test valve will be vibrated at the indicated
resonant conditions for the period shown in Table 1. and with the
amplitudes of vibratory accelerations as in THG.1614.

If more than four resonances are encountered with vibration applied along
any one axis, the four most severe resonances shall be chosen for test.

TABLE 1.

Number of Resonances.	0	1	2	3	4
Total vibration time at resonance(30 minutes at each resonance)	-	30 Min	1 Hour	1½ Hour	2 Hour
Cycling Time (Vibration Endurance)	3 Hour	2½ Hour	2 Hour	1½ Hour	1 Hour

Note: Times shown refer to one axis of vibration.

6. Functional Tests.

After the vibration tests the unit will undergo the functional tests detailed
in Paras 2.2 and 2.3.

Fig. 12.8 Page from development test schedule

mally the inspectors are not production workers for the obvious reason that
their motives must not be confused with the need to deliver the system.

The inspectors must have instructions; they must know what to measure
and to what tolerances. Normally then, the Q and R department will
produce inspection instructions from the drawings they receive from the
designer and from the draughtsman. Because different manufacturing processes may be involved, the Prototype Shop Inspectors may require different
instructions from the Production Shop Inspectors. Fig. 12.9 shows a page
from a typical Production Test Schedule.

PRODUCTION   TEST   SCHEDULE

FOR

5. Functional Tests (Lucas Flow Rig)

The tests are to be carried out with a nominal 24 volt D.C. supply. All pressure
measurements are to be made using static pressure tappings.

With the valve set up having a downstream volume of 680 cu in   and the cock terminating
this volume set to give a flow of 20 lb/min at room temperature when the valve downstream
pressure is 65 p.s.i.g the performance is to be recorded under the following conditions:

5.1. Room Temperature Tests. (Datum Valve Removed).

The tests are to be carried out by applying a false datum pressure equivalent to
the valve inlet pressure.

5.1.1. Failure Test.

With the solenoid de-energised, i.e. valve closed, the valve inlet pressure
must be increased to 145 p.s.i.g. The solenoid is to be energised and the
pressure switch must operate at between 78 and 80 p.s.i.g to close the
valve head. The downstream pressure may overshoot to 145 p.s.i.g pressure
but must not exceed 80 p.s.i.g for more than 1.1 secs

5.1.2. Valve Head Tests (Minimum Operating Pressure).

Repeat tests 4.1 and 4.2.

5.2. Control Characteristic Test (With Datum Valve Re-Fitted).

5.2.1. Steady State Test.

Energise the solenoid and then increase the valve inlet pressure slowly
from 0 to 145 p.s.i.g and decrease slowly back to 0. Between 65 and
145 p.s.i.g the valve downstream pressure must be stable and lie between
60 and 70 p.s.i.g on both rising and falling inlet pressure.

Note: If the above test requirements are not met due to overshoots,
hysteresis etc, then the following procedure is to be carried out:

The unit is to be subjected to a 15 minute soak on the flow rig with
an inlet pressure of 145 p.s.i.g and a flow in excess of 20 lb/min
at a through air temperature of 160°C. At the end of the soak period,
the unit will be cycled open and closed 10 times.

After cycling the functional tests at high temperature(Para 6.)may be
carried out at this stage.

After the functional tests at high temperature, the unit is to be returned
to room temperature and tests 5.2 are to be carried out.

5.2.2. Slam Acceleration Test.

The solenoid must be energised and the valve inlet pressure increased
from 0 to 145 p.s.i.g at a rate of 45 p.s.i/sec. The same control
bandwidth of 60 to 70 p.s.i.g must be obtained. An overshoot is
allowable provided the pressure switch is not actuated. The unit must
stabilise within 5 seconds of applying maximum inlet pressure. This is
be measured on a suitable recorder.

PREPARED BY	(	1 6.8.68.19	ISSUE No.	8	9	10		
CHECKED BY	!	21.8.68?	ALT. No.	89010	39242	9D698		
PROVED BY		: : 19	ISSUE No.					
D.I.S.			ALT. No.					

Fig. 12.9 Part page of a typical production test schedule

Buying Instructions (document 6)

If a system is to be made, material must be bought. The material may be the metal which is to be cut; bought-out components, already manufactured; special tools; and even the paper on which instructions are written. Buying these raw materials is a skilled job for a specialist. The Buyer must know where he can buy cheaply, which suppliers control the quality of the materials they sell, which suppliers deliver on time, when it is reasonable to expect lower unit prices for larger orders etc., but nevertheless, he must be told what to buy. Because quantities and methods are different, the Prototype Shop may require different purchases from the Production Shop, but in either case interpretation of the drawings and the shop-loading plan will enable the appropriate information to be transmitted to the buyer. It is reasonable, too, to expect the buyer's knowledge to feed back information which could modify the buying instructions.

PROCESS LAYOUT							PART No.						SKETCH
DESCRIPTION													
MATL. SPEC.		MATL. REQD. EACH						CODE		U.O.M.		PAGE	
DRAWING No.													
		LAYOUT ISSUE					DEC.		LT.	S.O.S.	SIG.		OF PAGES
OP. No.	DEPT.	LAB.	M/C GROUP	TIME ALLOW MIN. EA.	SU. TIME HRS.	BOX No.	TOOLING & DESCRIPTION OF OPERATION						

Fig. 12.10 Part of a typical process schedule

Planning Instructions (document 7)

The Production Department Operatives may not have the skill to work directly from detail drawings and, in any case, it may not be convenient for them to do so. Consider a machined part; this part may be manufactured from bar by a sequence of operations such as (i) saw off an appropriate length of material, (ii) turn a suitable diameter on a lathe, (iii) mill to a further dimension etc. While a skilled man may appreciate that all these operations are necessary and must be carried out in a predetermined sequence, such analysis is not the function of the machinist. It is probable, too, that each machine in the sequence is operated by a different man. It is more convenient then, to list the operations so that each man is given precise information on, and only on, the operation for which he is responsible, the source of the material on which he will work and the destination of the parts when his operation is finished. Such information must clearly be formal and usually takes the form of a Process Schedule (for a single part) or an Assembly Schedule (for assembling parts into a sub-assembly or assembly). These schedules will be the instruction to the shop-floor worker and will contain required numbers off as well as a job description; sketches may be added to clarify instructions; permitted times may also be added to indicate the basis of payment to the operative. Figs. 12.10 and 12.11 show typical Process Schedules.

The Hardware (8)

The Prototype Shop makes a system in order that the methods of manufacture may be explored and a system may be tested. When successfully manufactured, therefore, the prototype system(s) will be sent to the Development Department for testing. When tests of a prototype show the system to be satisfactory, further systems will be made by production methods and because there will be differences between methods of prototype manufacture and methods of production manufacture, early production models must also be submitted for testing by the Development Department.

When considered satisfactory, systems made by the Production Department will be delivered to the customer. During service, faults will occur and these will be monitored, reported and possibly repaired by Service Engineers. Some faults may result from errors in design and formal feed-back methods from the Service Engineer to the designer must exist so that such errors may be put right by modifications to the designer's instructions. Some faults may result from bad materials or inspection and in these cases the Q and R department must be involved.

PLANNING LAYOUT

PART No.	ISSUE 4.	DATE 11.11.69	ISSUED BY E.V.M.	SHEET 1/4		
OP No	DESCRIPTION OF OPERATION	TIME ALLOWED (Mins. Each)	SETTING TIME (Hrs. per 100) Initial	Maint.	TOOLS & GAUGES	
	Alt No.89264. Drawing Issue to 4. (Incorporating Alt.89851 3803 001 105 Issue 3). N.B: This is an oxygen valve and all parts must be free of grease and dirt.					
10	ASSY. DEPT. Unwrap parts and lay out in preparation for Assembly.					
20	ASSY. DEPT. Stamp the following on nameplate 3803 001 202. Type No. 3803 001 000 Serial No.	3.00				
30	ASSY. DEPT. Remove the Orifice and Filter Assy. FRJ/SA/1297 and lightly brush inner threads with lacquer COM.9452, or COM.9481, leaving outer two threads free of lacquer. Then over threads fit a Gasket FRJ. 1087. Screw into the threaded bore of Housing 3803 001 200, and cure at 120°C for ½ hour or at room temperature for 6 to 8 hours, if using COM.9452. If Using COM.9481 cure at room temp. for 2 hours only. FLOW TEST. Fit adaptor and flow check with 100 p.s.i. nitrogen or clean dry air applied to inlet. Flow to be between 25 s.c.i.m. and 96 s.c.i.m. (.014 s.c.f.m. and .056 s.c.f.m.). Remove. (This Op. makes 3803 001 105 sub. assy. us. 3). assembly is not to be u...				Adaptor AJT.	

Fig. 12.11 Part of a typical process schedule

Service Manuals (document 9)

The Service Engineers will be expected to keep a system serviceable, either by preventive maintenance, repair or instructions which enable the customer to rectify faults. Working from drawings (and also with access to reports of manufacture and test) the service engineer will produce manuals which give all the service and repair information necessary.

Feedback (document 10)

We see from Fig. 12.1 that the feedback of information is as important as the feed forward. Often this feedback is informal but in some cases formal documents are essential. An important case is where drawings have already been used for manufacture or have been made available to the customer. Under such circumstances it is necessary to distinguish clearly between up-to-

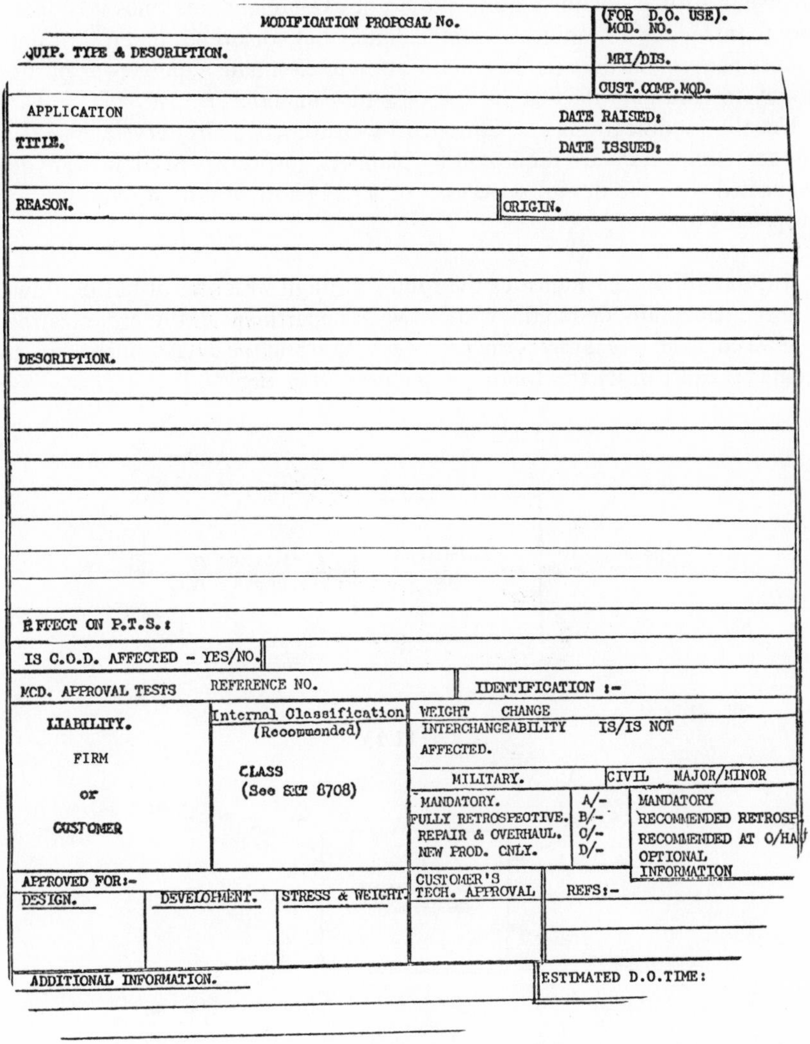

MODIFICATION PROPOSAL No.
(FOR D.O. USE). MOD. NO.
QUIP. TYPE & DESCRIPTION.
MRI/DIS.
CUST. COMP. MOD.
APPLICATION
DATE RAISED:
TITLE.
DATE ISSUED:
REASON.
ORIGIN.
DESCRIPTION.
EFFECT ON P.T.S.:
IS C.O.D. AFFECTED – YES/NO.
MOD. APPROVAL TESTS
REFERENCE NO.
IDENTIFICATION :–
LIABILITY.
FIRM
or
CUSTOMER
Internal Classification (Recommended)
CLASS (See EXT 8708)
WEIGHT CHANGE
INTERCHANGEABILITY IS/IS NOT AFFECTED.
MILITARY.
CIVIL MAJOR/MINOR
MANDATORY. A/–
FULLY RETROSPECTIVE. B/–
REPAIR & OVERHAUL. C/–
NEW PROD. ONLY. D/–
MANDATORY
RECOMMENDED RETROSP
RECOMMENDED AT O/HA
OPTIONAL
INFORMATION
APPROVED FOR:–
DESIGN.
DEVELOPMENT.
STRESS & WEIGHT.
CUSTOMER'S TECH. APPROVAL
REFS:–
ADDITIONAL INFORMATION.
ESTIMATED D.O.TIME:

Fig. 12.12 Part of a typical modification proposal form

date and out-of-date information so that every feed-forward document must carry an issue number. Modifications to drawings and other documents cost money, not only because of the actual work of modification but also because of the cost of manufacturing and testing labour that they initiate. In order to ensure that modifications are initiated only where faults justify it and also to ensure that faults are recorded (so as not to be repeated) and costs recovered where possible (if, for example, a change in the customer's requirements leads to modification, the customer should expect to pay for the work involved), formal documentation should be available. Sometimes it is necessary to introduce a modification immediately but sometimes it can wait until a more convenient time (is the modification to eliminate a dangerous part or simply to ease manufacture by use of a new material?).

Usually modifications are requested by customers or by service engineers but sometimes they are requested by planners, inspectors or almost anyone connected with the design process. Fig. 12.12 shows a typical Modification Proposal Form.

Although Fig. 12.1 indicates the routes through which information flows, it does incidentally define the work of each department in the organization.

If we consider as examples the prototype shop and the development department, we can extract the following pictures from Fig. 12.1.

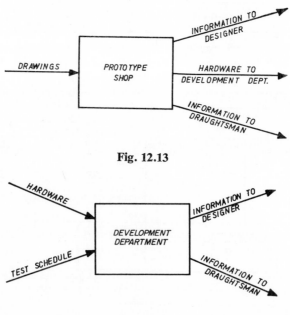

Fig. 12.13

Fig. 12.14

Although in each case it may be necessary for the department to communicate with people other than those shown, the outputs shown are the reasons for their existence. Any work that these departments do is valuable only if it contributes to achieving the objectives shown in the diagram. Playing the part of a business systems analyst, one can go through the entire diagram, Fig. 12.1, defining every job in the organization in terms of input and output. If one does this in a real situation one usually finds a lot of expensive, redundant work going on and a lot of necessary work not being done.

Only when the manufacturing and quality control instructions for the production process result in the supply of a system which meets the design specification, is the designer's job done. Eventually, that is after development, two sets of instructions will exist. The first set, consisting mainly of drawings, defines, in every detail, the geometry of the system and all materials. Such a set of information, in principle, would enable any manufacturing organization to make a satisfactory system. The designer is responsible for every aspect of this set of instructions.

The second set of information defines the manufacturing instructions in ways best suited to the resources of a particular manufacturer. The Planning Department is largely responsible for this set of instructions although the designer must take into account the manufacturing resources or the planners will have an unnecessarily difficult job and manufacture will be unnecessarily expensive.

Exercises and Subjects for Discussion on Part 3

1. Draw the design scheme of a system to meet the requirements of Exercise 5, page 49.

2. Draw the design scheme of a system to meet the requirements of Exercise 7, page 52.

3. In discussion groups, criticize the schemes of 1, 2 above from the point of view of:
   (i)   Prototype manufacture
   (ii) Production
   (iii) Development testing
   (iv) Service.

4. Draw the detail drawings and any other manufacturing instructions thought necessary for the Prototype Shop to make both of the systems proposed.
   (i)   Identify and discuss parts which could be standardized.
   (ii) Justify the tolerances on the drawings.
   (iii) Devise a production process schedule for one of the component parts.
   (iv) Criticize the methods and costs of manufacture of the part considered in (iii), above.
   (v)   Discuss inspection procedures for the part considered in (iii). Would inspection procedures be the same for production as for prototype manufacture?
   (vi) Identify and discuss a necessary inspection operation which is not a direct linear measurement of a dimension.

5. Draft a schedule of tests designed to prove that the systems meet their design specifications.
How would you prove the fatigue life of the motor car switch?

6. How would you number your drawings?
How would you incorporate necessary design modifications?
How would you list buying information?

7. What operating and servicing instructions would you issue in both cases?

Further Reading

T. Burnham and D. Bramley, *Factory Organization and Management*, Pitman, London, 1957.

T. Hicks, *Successful Engineering Management*, McGraw-Hill, New York, 1966.

Part 4

THE MANAGEMENT OF DESIGN

Part 4 deals with the control of work by network methods, the extension of network methods to resource allocation and controlling expenditure. Network-based management is justifiably fashionable because it is a powerful tool. The customer will want to see these methods applied to design just as they are applied to other projects. Design, however, involves the prediction of invention and development time and this raises difficulties when attempting to establish networks. There is also the problem that the simple Critical Path Methods usually taught to undergraduates ignore the biggest management problem of all, optimum use of resources.

There are many useful introductions to Network Based Management, among which may be mentioned Battersby, 1964, and Lockyer, 1964. Few texts deal with the problem of resource allocation although Buffa (15a) makes some reference to resource scheduling. Beyond the elementary level, the best texts are undoubtedly the manuals produced by the major computer companies to assist with the use of their packages. These packages usually deal not only with resource allocation but also with accounting. Information about the better commercial packages however, is often restricted to users of the computer for which the package was designed.

CHAPTER 13
Critical Path Methods

The designer has to design a system which can be delivered when it is wanted. Usually, in fact, he must predict the delivery date so he requires a means of predicting and a means of monitoring the time taken from the placing of an order to meeting the order. A useful tool in this work is the Critical Path Method.

Consider the much simplified design and manufacture process described in Fig. 13.1.

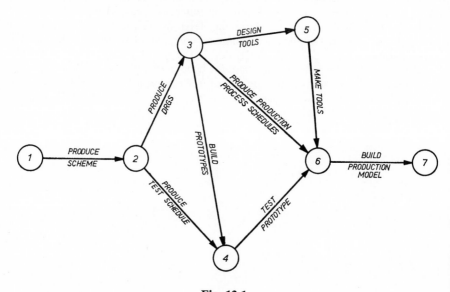

Fig. 13.1

Each numbered balloon denotes a milestone in the procedure while the lines denote activities which have to be done and which take time.

Activities may be sequential or parallel. Consider activity 3–4 (build prototype) and activity 4–6 (test prototype). These activities are sequential because 4–6 cannot be worked on until 3–4 is complete (i.e. you cannot test the prototype until you have built it).

Consider activity 3–6 (produce production process schedules) and activity 4–6 (test prototype). These activities are parallel because there is no reason why they should not be worked on simultaneously.

The milestones show which activities are sequential and which parallel and indeed tell us rather more. Activities 3–4 and 4–6 are sequential and so are activities 2–4 and 4–6 and the diagram shows that both 3–4 and 2–4 must be completed before 4–6 can start (i.e. testing cannot start until both the prototype is available and the test programme defined).

If the designer knows, or will predict, the time that each activity will take, he can put those times on the network thus:

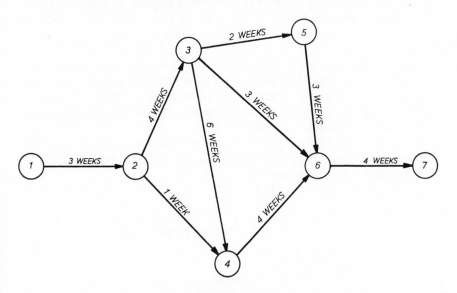

Fig. 13.2

This indicates his estimates that 1–2 (producing the scheme) will take 3 weeks, 2–3 (producing the drawings) will take 4 weeks, etc.

We may now move from milestone 1 to milestone 6 to calculate the earliest time at which any activity may be started and Fig. 13.3 is created.

If the job is started at week number 0, then milestone 2 is reached after 3 weeks so that the earliest starting date for activities 2–3 and 2–4 is week 3. If activity 2–3 is started at week 3 then milestone 3 cannot be reached before week 7. The earliest possible starting date for activity 3–5, activity 3–6 and activity 3–4 is thus week 7.

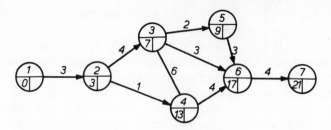

Fig. 13.3

Milestone 4 can be reached by two routes. One week after week 3, activity 2–4 may be completed but activity 3–4 cannot be completed until week 13, 6 weeks after week 7. Clearly, since both 3–4 and 2–4 must be completed before 4–6 can commence, the earliest starting date for activity 4–6 is week 13.

Similarly the earliest starting date for activity 6–7 is 17, the largest of 7+3 (arriving at milestone 6 by adding the activity time of 3–6 to the 7 at milestone 3), 13+4 (arriving at milestone 6 by adding the activity time of 4–6 to the 13 at milestone 4) and 9+3 (arriving at milestone 6 by adding the activity time of 5–6 to the 9 at milestone 5).

By this process, moving through the network, we can calculate all earliest starting dates and, in particular, the earliest date at which the project can be finished (week 21 at milestone 7).

The symbols used in Fig. 13.3 are:

Fig. 13.4

Knowing the earliest possible date at which the project may be finished, the designer may now calculate the latest permissible finishing date of each activity to produce Fig. 13.5.

Having shown that the earliest date at which the whole project can be completed is week 21, we wish to ensure that it is in fact, finished by that date.

Working back from milestone 7, reached at week 21, we know that the latest date at which activity 6–7 may be finished is week 21. In order to achieve this date, activity 6–7, which takes 4 weeks, must be started by week 17 at the latest so that the latest permissible finishing date of activities which imme-

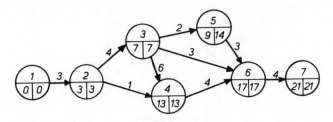

Fig. 13.5

diately precede 6–7 (or immediately precede milestone 6) is week 17. To ensure this, activity 3–5 must be finished by week 14 and activity 3–4 by week 13.

The latest permissible finishing date of the activity immediately preceding milestone 3 is week 7, the lowest value of 14 minus 2 (latest time at milestone 5 minus the time required for activity 3–5), 17 minus 3 (latest time at milestone 6 minus the time required for activity 3–6) and 13 minus 6 (latest time at milestone 4 minus the time required for activity 3–4). Proceeding in this way, we can obtain the latest permissible finishing date for the preceding activities at each milestone.

The symbols used in Fig. 13.5 are (Fig. 13.6):

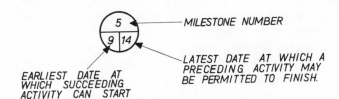

Fig. 13.6

Clearly if at any milestone the earliest date at which a succeeding activity can be started is also the latest date that we are permitted to finish a preceding activity, there is no time to spare. It we join such milestones along the activities which have no spare time we have the critical path of Fig. 13.7.

If any activity lying on the critical path takes longer to do than has been predicted, then the overall time for the whole job will be extended. The activities which do not lie on the critical path may take longer than predicted without affecting the finishing date of the whole job.

Consider activity 2–4. It is possible to start this activity at week 3 but as long as it is finished by week 13, the finishing date for the whole project will not be affected. Activity 2–4 is expected to take 1 week but could take 10

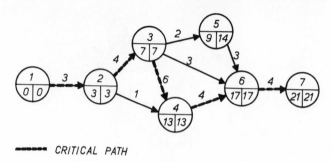

■■■■■ CRITICAL PATH

Fig. 13.7

weeks without affecting the finishing date for the whole project, so that this activity has 9 weeks total slack (or float) and the manager need not worry about it until it is very late indeed.

Similarly job 3–5 can start at week 7, must be finished by week 14, only takes 2 weeks and therefore has 5 weeks total slack. Job 5–6 must be finished by week 17, can start at week 9, takes only 3 weeks and therefore has 5 weeks total slack. Clearly job 3–5 and job 5–6 cannot both take 5 weeks longer than predicted because they will share 5 weeks slack and any extra time taken by one activity reduces the extra time available to the other.

Most jobs can be modelled by means of networks and the finishing date for the whole job predicted. The network has more use, however, than simply to predict the finishing date. Knowledge of the activities on the critical path tells the manager which activities must be rigorously controlled to the predicted time and hence where to direct resources when programmes are not being met. The activities that are not on the critical path can frequently be given second place for consideration by management and, more generally, the effect of any activity delay on the overall programme can be calculated.

The example used to demonstrate the Critical Path Method is so simple that its management would hardly require the use of special techniques. In real life, however, a network may have many thousands of activities and only by building such a network could one analyse the situation and calculate the delivery date. Networks in real life cannot reasonably be calculated by manual techniques.

Luckily, calculations of critical paths and slacks require only simple arithmetic (earliest starting dates plus activity times or latest finishing dates minus activity times) and logic (selecting the latest of the earliest starting dates or the earliest of the latest finishing dates) and such operations are easily done by a digital computer. It is comparatively straightforward to write a computer program which will make these calculations, but because of the extent to which Critical Path Methods are used, most large computer firms have programs already written and available. These programs are nearly universal in character and will fit almost any network. The input data cannot, of course, be in the form of a diagram but will consist of a list of all activities, the time of each activity, and for each activity a list of those activities which necessarily precede.

The nomenclature used above has been developed intuitively as the network of the example has been developed. No attempt has been made to define terms precisely although formal definitions are desirable. It is also desirable to create further parameters for use in more complex situations.

We therefore define:

Earliest Starting Date (ES) of an Activity

This is the earliest date at which work on an activity can start if we assume that no time has been lost on preceding activities.

For example, in Figure 13.2 the earliest starting date of activity 3–5 is 7; the earliest starting date of activity 5–6 is 9.

The ES for each activity has been calculated to provide data for Fig. 13.3.

Latest Finishing Date (LF) of an Activity

This is the latest date at which work on an activity may be permitted to finish if the whole project is to be completed as early as possible.

For example, in Fig. 13.2 the latest finishing date of activity 3–6 is 17; the latest finishing date of activity 3–5 is 14.

The LF has for each activity been calculated to provide data for Fig. 13.5.

We also introduce and define:

The Earliest Finishing Date (EF) of an Activity

This is the earliest date at which work on an activity can be finished.
If T is the time of an activity then

$$EF = ES + T.$$

and

The Latest Starting Date (LS) of an Activity

This is the latest date at which work on an activity may be started if it is to be finished by its LF.

$$LS = LF - T.$$

Total slack has already been discussed in the example and we may define:

Total Slack (TS) for an Activity

This is the amount of time that an activity could be permitted to overrun its predicted time when all preceding activities are completed as early as possible and all succeeding activities start as late as possible. It is the amount of time that *one* activity could be lengthened or delayed without delaying the completion date of the whole project.

$$TS = LF - ES - T$$

$$= LF - EF$$

$$= LS - ES.$$

It is also useful to introduce and define:

Free Slack (FS) of an Activity

This is the amount of time that an activity could be permitted to overrun its predicted time if all preceding activities are completed as soon as possible and all succeeding activities start as soon as possible. Free slack is the amount of time that an activity may be delayed without delaying subsequent activities and its importance arises from the fact that an activity with free slack may be delayed without rescheduling any other activity.

$FS = ES$ of next activity $- EF$ of this activity (note that free slack robs no subsequent activity of slack).

TS and FS are commonly used. Less commonly used are:

Interfering Slack

Interfering slack $= LF$ of this activity $- ES$ of next activity (note that interfering slack robs no preceding activity of slack).

Independent Slack

Independent slack $= ES$ of next activity $- LF$ of previous activity $- T$ (independent slack robs neither preceding nor subsequent activities of slack.)

Further useful additions to the repertoire are:

Dummy Activity

It is sometimes convenient to introduce a dummy activity which is quite artificial in the sense that it represents no actual activity and takes no time. A dummy activity is used for two reasons:
    (a) to denote a necessary precedence

(b) to avoid confusion when an activity is denoted by its preceding and succeeding events.

As an example of the first reason, consider the following simple network (Fig. 13.8):

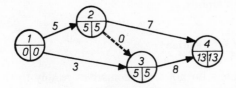

Fig. 13.8

Activity 2–3 is a dummy and shown dotted, as is usual with dummy activities. The dummy ensures that activity 3–4 may not start until activity 1–2 is completed.

In a computer program, the logic could be written without using a dummy activity but since most networks are drawn before being computed the dummy is invariably used where it conveniently expresses precedence rules.

As an example of the second reason for using dummy activities, consider Fig. 13.9:

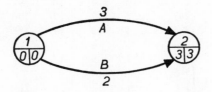

Fig. 13.9

Most computer programs would refer to both activity A and activity B as 1–2. Although the more sophisticated, commercially available programs do permit us to use suffices (e.g. to call up activity 1–2, A and activity 1–2, B) in order to avoid confusion, it is much simpler to use the following configuration (see Fig. 13.10).

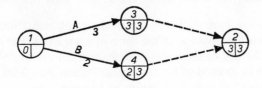

Fig. 13.10

In this configuration activity A may be called 1–3 and activity B may be called 1–4, without confusion by adding dummies 3–2 and 4–2.

Note that some critical path methods require, with any activity, that the succeeding event must have a higher number than the preceding event. This is not true of the more sophisticated packages (e.g. I.C.L. PERT) nor neces·sarily of manual methods.

Tabulated Output Information

Because computers do not lend themselves readily to the production of diagrams, data is often prepared and calculated in tabular form so that the network of Fig. 13.7 might well be described by the following table:

	Activity	Preceded by	T	ES	LF	EF	LS	TS	FS
*	1–2	—	3	0	3	3	0	0	0
*	2–3	1–2	4	3	7	7	3	0	0
	2–4	1–2	1	3	13	4	12	9	9
*	3–4	2–3	6	7	13	13	7	0	0
	3–5	2–3	2	7	14	9	12	5	0
	3–6	2–3	3	7	17	10	14	7	7
*	4–6	⎰3–4 ⎱2–4	4	13	17	17	13	0	0
	5–6	3–5	3	9	17	12	14	5	5
*	6–7	⎧5–6 ⎨3–6 ⎩4–6	4	17	21	21	17	0	0

* Denotes activity on the critical path

A table such as that above is often presented to the manager at regular intervals so that he can assess and control progress on a project.

Bar Charts (sometimes called Gantt charts after their inventor)

Whether Critical Path Methods are used or not, many managers find it helpful to present information in the form of a bar chart. Bar charts may be used to show the use of resources (see chapter 14) but are also frequently used as a different method of presenting the information from a network.

The information of Fig. 13.7 could be presented as the bar chart of Fig. 13.11.

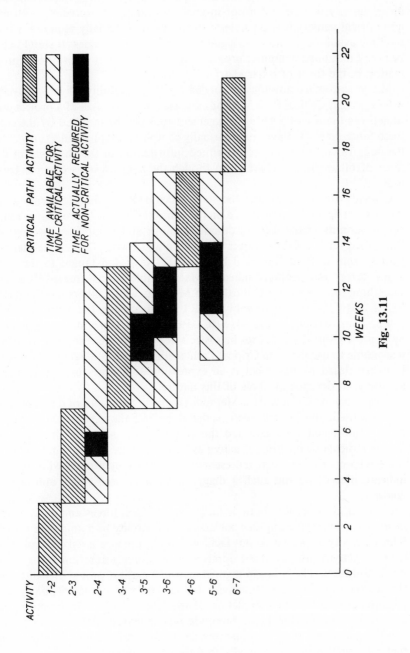

Fig. 13.11

It will be observed that Fig. 13.11 gives a very good pictorial representation of an activity in time but a not-so-good indication of precedence. With a non-critical activity it is possible to move the time actually required within the time available to show the significance of the slack times. It should also be noted that most computer programs have the capacity to print-out information in the form of a bar chart.

Many customers nowadays, worried by the possibility of late delivery, will insist that Critical Path Methods are used in the control of a project and sometimes the use of such a management tool will be required by the contract. Since network diagrams are usually complex and (physically) unwieldy, the design manager may present edited information to the customer in the form of tables or bar charts even though the network is used as the basic tool.

Critical Path Methods have shown so much power in the last twenty years that a great deal has been written on the subject. This has led to a large number of variants of the method described above and to at least two names being used for it. PERT (Project Evaluation and Review Technique) derives mainly from methods invented by the U.S. Navy and taught to the U.S. armed forces and weapons industry. Many people seem to regard PERT as something different from Critical Path Methods whilst others use the terms PERT and CPM interchangeably. PERT is also used as the name of a very comprehensive computer package (ICL PERT) which, in addition to analysis by Critical Path Method, uses heuristic rules to allocate resources. It seems reasonable to use the term Critical Path Methods as a general one, although Network-Based Management is an expression sometimes used, with more accuracy, to describe all tools of this nature.

Early work on Critical Path Methods threw up two types of diagram. The one now used, almost exclusively, is that described above, in which the activity is represented by a line and the events, at the nodes of the network, merely indicate completion of activities. Another type of diagram similar in some respects to Fig. 12.1, represented an activity by a node with lines to indicate precedence but such a diagram is rarely used today and is best ignored.

One feature of PERT, both as taught to the U.S. forces and as available in some commercial computer packages, is the ability to manipulate probabilistic activity durations. If this facility is used, someone is asked to predict, for each activity, an optimistic duration, a most likely duration and a pessimistic duration. The solution algorithm will then fit a probability distribution to the duration of each activity and calculate an optimistic, likely and pessimistic date for the completion of the whole project. In practice, few industrial users of Critical Path Methods use this refinement because although it was designed to cope with inaccuracies of forecasting, it is usually found that a foreman who has difficulty in forecasting the duration of an activity has even greater difficulty in making three forecasts and, in general, the results achieved do not justify the extra complication.

Most industrial users of Critical Path Methods find that the limitations of the simple methods described above are not overcome by the use of more advanced mathematical techniques. The problem of size has already been mentioned as one that is readily solved by the use of a computer, which can deal with thousands of activities, without difficulty. Less easily solved, however, are the problems which arise from the limitations of resources, complexities of calendars and cost control. Only when these problems are overcome are Critical Path Methods really useful in most industrial situations. Heuristic methods, normally requiring expensive program packages, are available for their solution and some of these will be touched on in the next chapter.

CHAPTER 14
Limited Resources

Although Critical Path Methods are very powerful, they take little or no account of the availability of resources. Resources may be money, men, machines or any class of item which is necessary in the design and manufacturing process and any of which may be in short supply (limited resources). In the simple examples which follow, the limited resources will invariably be men. While it may be feasible, technically, for two activities to be worked on in parallel, the resources available may just not permit this. It may be technically feasible to build the back axle of your car and put new piston rings in the engine at the same time but if there is only one man available to do both jobs then they will have to be done one after the other. In almost all real-life situations, limitations of resources rather than the simple critical path will dictate the completion date of a project.

Suppose that we have a very simple situation in which a team of one technical man and two craftsmen are completely responsible for a small project and let us suppose the project to be described by the network of Fig. 14.1.

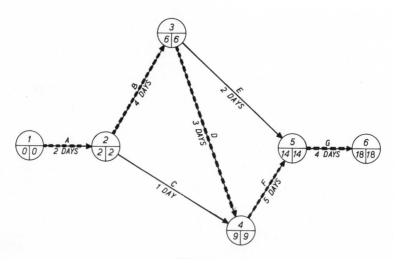

Fig. 14.1

Let us assume, however, that Fig. 14.1 does not tell the whole story and that we have the further information:

A, activity 1–2, requires 1 technical man
B, activity 2–3, requires 1 technical man
C, activity 2–4, requires 1 technical man
D, activity 3–4, requires 2 craftsmen
E, activity 3–5, requires 1 technical man
F, activity 4–5, requires 1 technical man
G, activity 5–6, requires 1 craftsman.

In Fig. 14.1, the network has already been considered by Critical Path Methods and appropriate dates calculated with no consideration of the limited manpower.

Let us also construct a bar chart (Fig. 14.2) to show a possible use of labour:

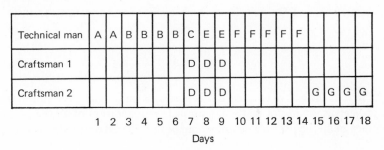

Technical man	A	A	B	B	B	B	C	E	E	F	F	F	F	F				
Craftsman 1							D	D	D									
Craftsman 2							D	D	D					G	G	G	G	

1 2 3 4 5 6 7 8 9 10 11 12 13 14 15 16 17 18

Days

Fig. 14.2

This bar chart indicates that we have allocated activity A(1–2) to our technical man for days 1 and 2; we have allocated activity D(3–4) to our craftsmen on days 7, 8 and 9 etc. Blank days in the bar chart indicate that our men are idle.

If we allocate the work in this way, we do in fact, complete the whole project in the time calculated by Critical Path Methods but we have used no real system in loading the work. When activity A is completed, the network tells us that both activities B and C can be worked on. In fact, because there is only one technical man, either B or C must be worked on first and the Critical Path Method does not tell us which to choose. If activity C were given priority over activity B, then our bar chart would have been that of Fig. 14.3.

In this case the project takes a day longer to complete than we calculated by Critical Path Methods. There are two, possibly, good reasons for loading activity B before activity C:

in the first place, the whole project is trivial and there is no difficulty in calculating the complete consequences of both courses of action (i.e. B before C or C before B),

in the second place, it would seem reasonable to load activity B before activity C because activity B is on the critical path and we know that any delay in doing an activity on the critical path must lead to a late finish for the whole project whereas we know that activity C has a float of 6 days.

Technical man	A	A	C	B	B	B	B	E	E		F	F	F	F	F				
Craftsman 1								D	D	D									
Craftsman 2								D	D	D					G	G	G	G

1 2 3 4 5 6 7 8 9 10 11 12 13 14 15 16 17 18 19

Days

Fig. 14.3

The first of the above two reasons is perfectly valid in all cases but in most real projects it could not be used. Complete enumeration of all the possible outcomes of all the possible decisions can be considered only with very simple

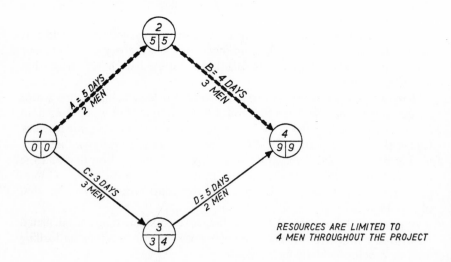

RESOURCES ARE LIMITED TO
4 MEN THROUGHOUT THE PROJECT

Fig. 14.4

projects. With the sort of project that would occur in real life, complete enumeration would take too long and cost too much to compute.

The second of the above reasons (i.e. choose to load an activity on the critical path rather than an activity not on the critical path) is not generally valid. Consider the simple situation of Fig. 14.4.

If we adopt the policy of loading activities on the critical path, in preference to those not on the critical path, we would start by loading A before C. We cannot load these two activities at the same time because we have insufficient men to do both jobs at once. When A is finished we can consider loading either B or C but B would take priority because it lies on the critical path. Following this procedure our bar chart becomes Fig. 14.5:

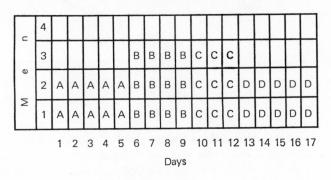

Fig. 14.5

It is easy to see that there is a better way of doing the work. Consider for example, the bar chart of Fig. 14.6.

This procedure completes the whole project 5 days earlier than that of Fig. 14.5, although C has been loaded before A.

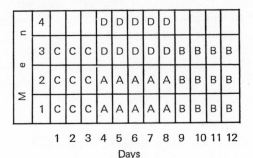

Fig. 14.6

The idea of loading a critical activity before a non-critical is not only a rule which cannot be trusted, it is a rule which cannot necessarily be applied. Consider the following situation:

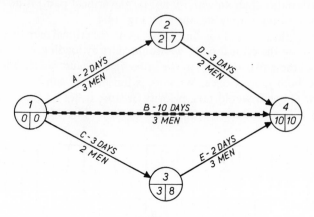

Fig. 14.7

If we give priority to critical activities then we will load B first but after loading B we have a choice between A and C. Since neither A nor C is critical, our rule is of no assistance. If we are to adopt rules for scheduling jobs then we must ensure that the rules really do generate decisions. The usual way of dealing with a situation in which a rule is inadequate, is to make a list of rules which can be applied in a stated order and our final rule will ensure that a selection is made, even if we simply select the activity with the lowest identifying number.

Apart from complete enumeration, which is not normally practicable, there is no procedure which we can use to schedule work and which ensures that we do so in the best way possible. Any algorithm which we use to schedule work will therefore make use of arbitrary* rules, which may seem reasonable but which will not necessarily give the best possible results. Most large computer companies have programs available for work scheduling and most of these programs appear to give better results than would most attempts to schedule a project manually. These programs usually make use of lists of priority rules and the rules may be quite complex, even though they are arbitrary.

---

* The reader may quarrel with the word arbitrary. The rules are arbitrary in the sense that there is no analytical justification for their use. The rules are selected because they seem, intuitively, to be likely to give better results than would random scheduling. In fact, research has shown that a computer program based on simple, reasonable rules gives very much better results than a computer program using random scheduling.

Although there is normally not much profit in using simple rules, manually, we can consider the project of Fig. 14.4 as an example of the method. Let our priority key be
  1. Earliest LF
  2. Earliest EF
  3. Least TS
  4. Alphabetical order.

From the network diagram we may immediately write down the following table:

		Latest finish	Earliest finish	Total float
2	A	5	5	0
4	B	9	9	0
1	C	4	3	1
3	D	9	8	1

We know, from our network, that either A or C must be loaded first and, using our key, the first rule tells us to load C (earliest LF), thus (Fig. 14.8):

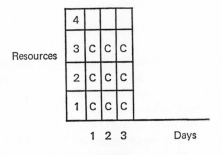

Fig. 14.8

Having loaded C, it is technically feasible to load either A or D but our first rule selects A and our bar chart becomes that of Fig. 14.9.

Our next choice lies between B and D and since both these activities have the same earliest finish dates we have to use rule 2 which selects D because this activity has an earlier EF date than B. D can commence on day 4 because there are enough men to work on A and D and our only other possible difficulty (technical feasibility) is overcome because C has been completed by day 4. Our final bar chart is thus that of Fig. 14.10:

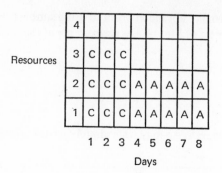

Fig. 14.9

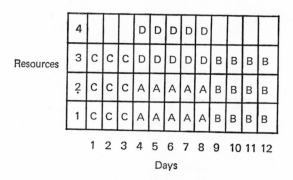

Fig. 14.10

It must be stressed that it is really fortuitous that the priority rules in the example give the best loading. Generally, the best that can be hoped for is that a reasonable choice of priority rules will give a good loading.

Normally, anyone who has to manage a complex design project would expect to use a computer program and he would hope that such a program would be already available from the large computer companies. The priority rules available to him would be much more sophisticated than those discussed above and several trial runs would assure the manager that he had selected reasonable priority rules.

If we study the commercially available programs for analysing networks and allocating resources we find that the following extensions are often made to the simple Critical Path Methods:

Continuous Feed Activities

If we consider an activity like drawing, we realize that it is not essential for all drawings to be finished before any prototype manufacture starts. One way of coping with such a situation is to introduce lead and lag activities as in Fig. 14.11.

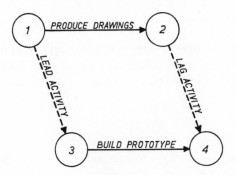

Fig. 14.11

1–3 is a lead activity and would have a duration equal to the time required to produce the first usable drawing. 2–4 is a lag activity and would have a duration equal to the time required to complete manufacture after the last drawing had been issued. Lag time would probably be the time taken to make the last detail and assemble and inspect the system.

Splittable Activities

If we consider activities such as drawing or calculating we see that they are splittable in the sense that if a draughtsman or stressman were interrupted before all his work on an activity were finished, the work done before the interruption would not be wasted. If, for example, a draughtsman is asked to draw two details and is taken off the job when he has completed only one, that one completed detail is useful and the man's time has not been wasted.

If, however, we consider an automatic lathe which is being set up for a complex machining operation and we stop the job before we have completed setting it up in order to use the lathe for another more urgent job, then the work done before the change-over is wasted. In this sense the setting up for the first job was a non-splittable activity.

A more homely example would be the painting of a door. Putting the top coat on would be a non-splittable activity—at least if we wanted the final effect to be reasonably good.

Constant Total Resources

Some activities will require a fixed amount of resources, however long they take. A component which requires to be machined from a pound of stainless steel will take a pound of stainless steel regardless of how long the machining takes.

Constant Rate Resources

Some activities use more resources or fewer, according to the time taken. Machining a pound of stainless steel will not use more steel if the machining takes twice as long as predicted but it will use a lathe at a constant rate for twice as long.

Calendars

In any real situation we have to contend with the fact that resources are not available at a constant rate, however they may be required. If the factory is working a five day week then Saturday and Sunday may have to be eliminated from our working calendar as far as human resources are concerned. Holidays may also have to be eliminated from the calendar or, if summer holidays overlap, may create periods in which the levels of available resources are lowered.

Threshold Resources

Overtime work at a high rate of pay may mean that men would be available outside normal working hours, but we may prefer not to use them unless compelled to do so. For every day in the calendar then, we would have a normal resource level and a threshold resource level, which we will use only if the 'due date' demands it.

One method of coping with a resource-limited situation is to:
1. Use the simple critical path method, ignoring resources, to calculate ES, LS, EF, LF, TS and FS.
2. Consider each period in the calendar in turn. At each period:
   2.1. List activities which are feasible from a study of the network only.
   2.2. Arrange those activities in an order of priority by some rules such as those discussed on page 123.
   2.3. Consider each feasible activity, in order of priority and load for the period under consideration, if resources are available (as resources are allocated they cease, of course, to be available for an activity of lower priority).
As activities are completed they are eliminated from further consideration. The procedure for making the decision of 2.3 seems simple but when complex resources and calendars are being considered, and when late delivery is likely, it may require a complex set of rules to decide whether to use thres-

hold resources or not; whether to split an activity or not. The set of rules used may depend on whether time is more important than economic use of resources or *vice versa*, and many sets of decision rules are available for the user of a typical computer software package.

Once again the decision rules will be arbitrary and often it will be necessary to test one set of rules against another to see which is better for a particular user.

Even in the elementary cases, considered above, we can see:

In a resource-limited situation, it is not generally possible to meet the delivery date suggested by simple Critical Path Methods.

However well a project is scheduled to the resources, the result will be (at least theoretically) that a lot of the resources will be idle for a lot of the time.

This latter point is fairly commonly appreciated even by people who do not use quantitative techniques for scheduling work. Most organizations go some way to reducing idle time by working on several projects at once and this naturally tends to reduce idle time but it also makes the work-loading problem much more complex.

Normally a designer will be dealing with many projects at the same time. Part of the design team may be discussing a new project with a customer, in the hopes of getting a new order, part of the team may be inventing and analysing possible design solutions to a recently contracted problem while yet other parts of the team are developing, manufacturing or collecting service information on earlier projects. Even where the designer is not managing a large number of different orders, he will be managing the designs of a large number of parts of a large project and these parts must be phased in such a way as to keep everybody busy all the time.

Having good ideas is not a quality which, by itself, will make a good designer. Good ideas are essential to good design but unless they are managed properly they may simply peter out in a mass of bad details or wasted time. Unless the designer can discipline himself and his colleagues to avoid both unnecessary work and unnecessary idleness, his products will not be competitive in price. This discipline can rarely be achieved in a complex situation unless use is made of management tools to understand and control the work. These management tools are, themselves, becoming a specialized branch of knowledge and in many companies it has been found desirable to employ specialists to design and operate management methods.

The examples used above to illustrate the problem of limited resources consider only one or two types of resources but it will be realized that in manpower alone, the design project may make use of performance engineers, designers, draughtsmen, development engineers, skilled fitters, skilled turners, semi-skilled fitters, semi-skilled turners and many others, while material resources are too many and varied to enumerate.

Any really useful scheduling program must take account of all such complications of this type and writing and operating such programs may be a job for a specialist in management tools but the designer must understand, broadly, how to manage a design project because without that understanding it will not be possible for him to make sensible design decisions where cost and time are involved. Even where a management services team exists to operate the management tools, design decisions are the responsibility of the designer and he must understand the meaning of management information even if it is someone else's responsibility to generate it.

CHAPTER 15

Accounting

If you invest your money in Savings Bonds it will earn you more than 7% per year; investing in a local authority will bring in over 9% per annum and these incomes will be virtually without risk. Engineering is a risky business, as a glance at any copy of the *Times Business News* will show, so that anyone who invests his money in it will expect a much higher return on capital than the 9% he can get without risk. Borrowing money costs money. Depending on where you borrow it from, borrowing money may cost from 8% to more than 25% per annum and if you have no money of your own to put into an engineering business, you will have to borrow it at say 10%. Clearly, if our business is run on borrowed money we do not start making a profit for ourselves until we have paid the interest.

Whether we have borrowed money for our business or whether we have used our own (as shareholders if not outright owners) we will clearly need to make an annual return on capital of the order of 20% to justify the work done and a return of about 10% will do no more than maintain the value of the money invested.

If we wish to judge the value of our work we must really compare its return with that obtainable from riskless investment. Suppose we could invest £1000 without risk, in a local authority at 10% then after 1 year we would have £1000 × (1·1) = £1100.

after 2 years we would have £1000 × (1·1)^2 = £1210
after 3 years we would have £1000 × (1·1)^3 = £1331

....

and   after n years we would have £1000 × (1·1)^n

so that the money which has a calculable, nominal value varying with time has a present value, P_v, of £1000.

If we wish to judge the merits of investing £1000 in a risky engineering venture in the hope of getting back £1500 in three years and if we believed that a riskless investment could be made at 10%, we would have to compare our £1500 not with the original £1000 but with the £1331 that riskless investment would have earned. Normally we do not compare earnings at some future date but, for the sake of using a consistent yardstick, we compare present values. Thus if we again consider 10% to be a reasonable risk-free investment rate, £1500 in three years would have a present value of

$$P_v = \frac{£1500}{(1\cdot1)^3} = £1127$$

We do not normally have to work out present values in this way because tables have been calculated for us (15a). Often too, we do not compare present values in precisely this way as it is sometimes convenient to compare constant annual returns on the basis of the annuity needed to buy those annual returns.

Normally this calculation of discounted cash flow would be of only indirect interest to the designer but the above discussions indicate that a designer is not worth his keep unless his efforts lead to a return on capital invested, of the order of 20%. We will see later that another way of looking at this is to expect a man to create riches to the extent of about three times his salary every year. We also see that the longer the period between starting and finishing a job, the more profit it must show. £1000 income from a sale three years in the future has a present value of about £750 whereas £1000 from a sale next year has a present value of about £910.

We have already seen, in earlier chapters, that the required delivery date imposes a constraint on the designer but now we see that, even apart from any required delivery date, time has a money value.

The more sophisticated the product the more will design costs have contributed to the total cost of the product and when the designer contracts with the customer to supply a system to meet the customer's needs, he usually contracts to do so by a certain date and for an agreed price. The designer has, therefore, to predict the cost and time of design and manufacture. Generally, when the geometry of a system and its parts is known and any special manufacturing processes specified, the costs of manufacture can be predicted without much error, but technical processes, invention, drawing, development, rectifying faults are all processes for which it is difficult to predict the amount of effort that will be required.

Much technical work consists of solving problems and it is always difficult to predict how long it will take to solve a problem (how long will it take you to solve a clue in the *Times* crossword puzzle?). Much technical work too, consists of rectifying faults and predicting the time required to do this would presuppose a knowledge of the faults in advance. If one could predict faults in advance, one would not allow them to happen.

Solving clues in the *Times* crossword may be an apt analogy because anyone with experience of such puzzles knows that sometimes you know the answer to a clue as soon as you have read it whereas on other occasions it will take hours (or days) to think of the answer. Usually, however, if you are experienced, you can form an idea of how long it will take you to do the whole puzzle on most occasions. This is not done by looking at each of the thirty or so clues and predicting the time that it will take to solve each one and then summing those times, but by an accumulation of experience about how long whole puzzles have taken to solve in the past. In industry, designers have too commonly predicted delivery dates and costs by breaking down large projects into small activities, studying each small problem in detail, predicting the required time and cost for the solution of each small problem and building the cost and time for the whole project from these predictions. This usually gives very poor estimates of costs and delivery dates and indeed makes it possible for the designer to justify almost any predicted cost. If we look at a very large number of small projects and assume that each will bring a difficult (but as yet unspecified) manufacturing or development problem then an estimate of the time to solve every problem will be pessimistic. If we assume that every small project goes through the system smoothly and without unforecast failures then our estimate of time will be optimistic. The optimistic and pessimistic answers usually differ by orders of magnitude and the designer is in a dilemma. If he quotes a high price to ensure a profit, however badly things go, his price will be too high and his competitors will drive him out of business. If he quotes a low price to make sure of getting the order then he may well find that only a few difficulties will result in loss of money and late delivery. Since the quotation is made when the designer is more preoccupied with getting an order than making a loss, there is often a tendency to underestimate costs. Particularly in the fields of weapons, aircraft and other large projects where governments are the customers, the final price for the system turns out to be many times the original estimate. Many people believe that the underestimates are simply due to lack of interest in money on the part of government departments and cynicism on the part of suppliers. In fact there is very little reason to expect a designer to be able to predict the delivery date and cost of a not yet invented system, particularly if the system is likely to require the development of new areas of knowledge.

Although the problem is difficult and not necessarily capable of solution, the designer nevertheless has responsibility for at least attempting to predict and control cost and delivery date. Perhaps because they have been much criticized, the industries with the most sophisticated design problems, the aircraft and weapons systems firms, are very cost-conscious and have been responsible for the development of most of the modern management tools, now used so successfully in most large commercial organizations.

If the designer is to have any success at all in predicting costs, he must accumulate experience of how much similar jobs have cost in the past. This means keeping records and, because most of the cost of any project is payment for labour, the records must include the man hours of different classes of labour that have been found necessary for every activity tackled in the past. Such records can form the basis of future estimates but they have other uses.

Most companies ask the workmen to account for their time. If the designer, draughtsmen, development engineers, etc. record, weekly, the ways in which they have spent their time it is possible to build up a picture, after the event, of most of the costs associated with a project. Some men cannot account for their time easily, however. The Personnel Manager may be necessary to a factory but he cannot record having spent his time in so many hours on one project, so many on another. Even the Chief Designer may spread his time thinly over a large number of projects and a number of administrative issues. Such indirectly accounted labour must be allowed for in every consideration of costs even if only by the simple process of spreading its cost over all projects evenly. Other costs too are involved. The development engineers' laboratory, the craftsmen's machine tools, the building they all work in, the heating, lighting, holidays, sickness, etc. are all examples of costs which have to be met and which, in the end, the customer pays for. The usual method of dealing with such indirect costs is to lump them all together as overheads so that were a draughtsman to be paid say 85p an hour, his time might be costed at, say, £2·50 an hour, the extra £1·65 being an allowance to cover such overheads as payment for administrative staff, capital equipment, heating, lighting, holidays etc. The ratio of indirectly accounted to direct labour costs in a modern dynamic industry is constantly changing and the need to know this ratio accurately provides us with yet another reason for keeping proper accounts of costs. If you think that your overheads amount to £1·65 an hour when they really are £1·85 an hour, you may find yourself quoting too low a price for profit. Knowledge of rising overheads is not only essential to enable proper prices to be quoted but may reasonably trigger off an investigation into the need for the overheads.

If we consider the above figures, we see that a man who is paid 85p per hour, who creates riches to the value of three times his wage, i.e. £2·55, will have produced a return of only 5p on an investment of £2·50.

Accountants frequently use procedures that are less sophisticated than is desirable. They may, for example, lump together too many costs as overheads and they may charge the same price for widely different grades of labour, but some method of accounting must be used and understood by the designer if he is to be aware of the likely costs of his decisions. The designer, who usually possesses more analytic skills than the accountant and more knowledge of computers, is frequently able nowadays to make a more detailed analysis of costs so that different charges may be made for different labour skills and overheads may be more correctly apportioned.

This keeping of records from time sheets thus enables the balance of labour to be checked, it enables the ratio of direct to indirect labour to be checked and it enables the cost per hour of direct labour to be calculated. It also builds up a library of knowledge of how long activities take and this library can help in the prediction of costs and times of future jobs. The designer will certainly be able to judge how long, on average, it takes to produce a drawing of a certain size, how long it takes to mount a 100 hr vibration test and many other fairly routine jobs. On the other hand only limited help can be obtained from such records when the work involved is finding and correcting unpredicted design faults.

Most new designs really reach an acceptable level very much by evolution. That is, over the years faults are evidenced and are eliminated, often by trial and error. Even the best designer will be unable to predict all the difficulties that he will encounter, he will merely predict more of the difficulties than a worse designer.

Another aspect of the designer's work is control of the project while it is in hand. When things do go wrong, the designer must know. While some expensive design and development troubles are caused by genuine lack of knowledge and can only be overcome at great cost, many problems can be reduced in size if they are appreciated in time and studied by the right man. There was a time when metal fatigue was not properly understood and one should not be overcritical of designers who did not foresee fatigue problems or solve them cheaply when they arose. There was a time when aerodynamically induced vibration was not understood and designers could not then be expected to forecast problems caused by this nor solve those problems cheaply when they occurred. There are, however, a host of problems which can be solved if the right knowledge is applied at the right time. Is the draughtsman spending more time than predicted on some detail problem? Perhaps the Chief Designer can solve the problem quickly or think of a gambit that removes the problem—provided that he is aware of it and its importance. Is the development engineer spending more time testing than was predicted? Perhaps the Chief Engineer can divert effort or test gear to reduce the time required—provided that he knows about the difficulty and cost of the testing.

We need then, a method of recording costs and of showing as soon as possible where predicted costs are likely to be exceeded.

One method of controlling costs is a development of Critical Path Methods. If a computer package is already being used to apply network-based methods then cost control is only a simple addition. Money is, after all, only another resource, and indeed, every other resource can be expressed in terms of money. Clearly, if we know what men are to be employed on any activity and we know the cost per hour of each man, it is simple arithmetic to calculate the cost of the labour for the activity. We should also know the costs of our materials so that knowledge of the materials required for each activity will lead directly to the total cost of materials. Overheads can also be calculated, for just as it is possible to introduce constant rate resources we can use constant rate costs. A shop overhead can be allowed for by associating, with that shop, a constant expense (or constant multiplier of labour and service costs) which will be applied for the whole time that an activity uses that shop.

With our Critical Path Methods and our records of work done we can fairly readily make three assessments:

(a) Before the first event, determine when each activity will be started and finished and estimate the resources and costs that will be used in every period until the project is completed.

(b) From time sheets and records of orders placed, determine how much work has been done, how much of each resource used and how much money spent at predetermined intervals during the project. Depending on the nature of the project, the Critical Path Program may load work hourly, daily or weekly. Records may show achievement week-by-week or month-by-month (rarely day-by-day since management decisions could not be taken that frequently).

(c) At each predetermined interval, repeat (a) but using as input data the current position shown by the records to recalculate start and finish times and resource and cost consumption.

The above three assessments can be further analysed to give at any review period during the project:

(d) how much work we expected to complete and
(e) how much we expected that work to cost $\Big\}$ from (a)

(f) how much we have actually done and
(g) how much money we have spent $\Big\}$ from (b)

(h) how much more we have to do and
(j) how much more money we expect to spend $\Big\}$ from (c)

We already know from (a):

(k) the total predicted cost.

Comparison of (d) and (f) will include a list of activities started but not finished after the anticipated time and so enables the designer to concentrate his attention on those activities which are not going well.

If we judge everything on a money basis we can say that, at a given review period, we expected to complete a fraction of the work equal to (e)/(k) whereas we have actually completed a fraction equal to (g)/[(g)+(j)], having spent a fraction of the intended expenditure equal to (g)/(k).

A possible improvement to this system is to repredict the anticipated duration of uncompleted activities at each review period. This is useful because as more of the project is attempted we learn more of the problems so that (h) and (j) may well justify computation on revised data.

The above serves not only to enable the designer to control the project but also to answer the question, perpetually posed by managing directors, 'What percentage of our proposed investment have we spent and what percentage of the work have we done?'

A less ambitious method of controlling projects (which nevertheless needs computer backing), particularly where a large number of small projects is concerned, is to work only from time sheets and orders for material. Each project is broken down only into the classes of labour which will be required and costs of material. Each labour group (e.g. Performance Office, Drawing Office, Prototype Shop, Development Dept.) is responsible

for estimating, before the start of work, how much work it will be required to do,

for completing weekly, time sheets showing how much time has been spent and

for showing on those time sheets, how much more work is yet to be done.

A draughtsman may, for example, be given 80 hours to produce some drawings and after working 20 hours will become aware of difficulties not previously anticipated. He could then indicate the problem by recording 20 hours of work completed but an anticipation of say, 100 hours of further work before the particular set of drawings would be completed. Such information, if quickly monitored, would indicate trouble to the designer. If all work were monitored in this way, the designer could quickly appreciate and deal with the most important (i.e. most expensive) problems. Top management would also benefit by such information because it could be made aware of progress, usually by showing the amount of money spent (both in absolute terms and as a percentage of that allowed) on a project and the achievement (usually as a percentage of the anticipated expenditure). If accounting methods are adequate, the senior managers of a factory can be made aware early that, whereas $x\%$ of the money available has been spent, only $y\%$ of the work has been done.

The two accounting methods discussed above do not differ much in principle although the second clearly ignores the feasibility network. Sometimes the information from an accounting system will only serve to make the designer and his employers wiser men although, at worst, it will enable the right problems to be investigated early and in some cases (as discussed in chapter 7) the information will enable the designer or manufacturer to negotiate a higher price with the customer.

Sophisticated accounting techniques for week-to-week monitoring of costs are often expensive to design and require either a computer or electronic data processing equipment to operate. It can take a year or so to program and debug a good accounting system although, once installed, such a system can be cheap to run. Some programs are available from computer firms but usually a lot of modification is required to tailor a program to the needs of a particular company.

Remember too, that too much information can be an embarrassment. In most companies too much is going on for any one man to be able to understand everything as it happens. A manager must give most of his consideration to those things which need it and little to the work which is proceeding reasonably according to plan. Any accounting system for the designer should therefore indicate quickly and easily where trouble lies and not bombard him with pounds of computer print-out about the normal day-to-day operation of a factory.

Exercises and Subjects for Discussion on Part 4

1. The activities of a project are related in the following way:

Activity A can start at week 0 and takes 8 weeks;
D can start at week 0 and takes 7 weeks;
B can start when A is complete and takes 12 weeks;
E can start when A and D are complete and takes 5 weeks;
C can start when B and E are complete and takes 8 weeks;
F can start when A and D are complete and takes 7 weeks.

Completion of both C and F completes the project.
Draw the network for the project and determine the critical path and the minimum time to completion.
Draw the bar chart for the project and mark clearly the Total float, Free Float and Interfering Float for activity D. How much Independent Float has activity F?
If Activity A requires 2 men and all other activities 1 man, plot the manpower histogram for the project. If only 2 men are available and either can work on any job, how should the work be arranged for the earliest possible completion date to be achieved? What are the implications of this arrangement?

2. A project involves the activities listed below. The length of each activity (in weeks) is given in brackets after the letter denoting the activity.

Activity	Constraints
A(3)	On letting of contract (at week 0)
B(4)	On letting of contract (at week 0)
C(6)	After completion of A
D(7)	After completion of A
E(5)	After completion of B
F(3)	After completion of C
G(9)	After completion of C
H(5)	After completion of D and E
J(4)	After completion of F
K(7)	After completion of G, D and E
L(5)	After completion of H
M(5)	After completion of J, K and L

Determine the minimum time to complete the project outlined above and which activities are on the critical path.
What is the Total Float of Activity F and what part, if any, of this is Interfering Float?

Draw a bar chart for the project outlined above. The manpower requirements for each activity are as follows: Activities A, B, C, D, E, G, J, L and M require one man, F and K require 2 men, H requires 3 men. Plot a histogram showing the total manpower required through the contract. If any man can perform any task, what is the minimum peak manpower needed if the contract is to be completed in the minimum possible time?

If you were the Chief Engineer with in-line responsibility for all the operations described in the project outlined above, how would you endeavour to ensure that it was completed to programme? In what way would you consider it prudent to modify your approach if the project were about a hundred times larger and occupied several years?

3. The manufacture of a new machine starts with a design study which takes 3 weeks and a preliminary market survey taking 5 weeks. Following the design study, detail design is started. This takes 6 weeks but some components require stress analysis lasting 2 weeks, followed by preliminary testing for 2 weeks, the stress analysis being started when the design study is complete.

Following completion of the detail design, manufacture of special jigs for mass production may start, provided that the market survey justifies continuance of the project and that the preliminary test results are satisfactory. Jig manufacture takes 4 weeks.

Comprehensive tests must follow the preliminary testing and this takes 6 weeks. Following these comprehensive tests and the manufacture of the jigs, mass production may start provided that material is available, the material being ordered when the design study is complete and delivery being expected to take 4 weeks. Production of a sufficient quantity for initial sales takes 6 weeks.

An advertising campaign is also started when the detail design and market survey are complete.

(a) Draw an activity/event network for the project and determine the critical path.

(b) Produce a chart which may be used to control the progress of the project.

(c) Determine the earliest date at which the advertising campaign can start and how long the campaign may last if it must be completed before the first batch of machines is ready for sale.

(d) Assuming that the delivery for material is known to take exactly 6 weeks is there a more sensible date at which to order the material?

4. A firm decides to market a new product as quickly as possible. Design takes 10 weeks and while it is proceeding, modification to a sub-assembly is carried out needing 4 weeks in the laboratory. With these modifications completed and the design finished, detailing can proceed for an estimated 4 weeks. Test specifications for the whole equipment can only be written when detailing is complete, whereas the test specifications for the sub-assembly can be prepared when the modifications are complete. Each set of specifications takes 3 weeks to write and when both are finished the necessary test equipment can be assembled, a job taking 7 weeks.

Manufacture of the whole equipment, estimated to take 15 weeks, can proceed after detailing and the product can be finally tested when it is made and the test equipment is available. This final test takes 4 weeks.

(a) Draw the network and determine the critical path.

(b) Draw the Gantt chart for the project and mark clearly on it the periods of time representing the Total Float, Free Float, Interfering Float and Independent Float for the preparation of sub-assembly test specifications.

(c) Each activity, other than the 15 weeks manufacturing period, requires one man. If the same man has to carry out all these activities (other than 15 weeks manufacture) how will the network and completion time be modified?

5. The figure shows two projects A and B, both of which use the same type of manpower resource (R). They can both commence at the same date, if necessary; assume

Project A

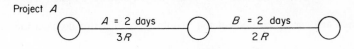

Project B

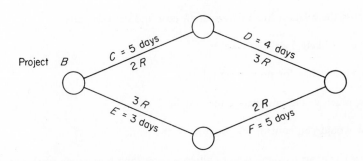

Resource limit = 5 R throughout project

in the first instance that activity A of project A is splittable. Schedule the projects to the following key:

1. Priority—project B has the highest
2. Progress
3. Latest finish
4. Earliest finish
5. Total float

To produce:

(a) Resource-Limited schedule
(b) Time-Limited schedule

In the case of (a) above if activity A is non-splittable how long will the projects take to complete?

For (b) above: if extra manpower can be obtained at a cost of £10/man-day, what is the extra cost of meeting the due date?

If all activities are non-splittable and if project A is given the highest priority, how long will the two projects take to complete if the resource availability is adhered to?

Question 6 can form the subject of several afternoons of group discussion.

6. Sketch and discuss typical activity networks of the work required to design and develop the devices proposed in exercises 7, 8 and 9, p. 76.

Put reasonable times to each activity and hence calculate the critical path, earliest and latest starting and finishing dates, total and free slacks for all activities.

Discuss the types of labour that would be required to do each job.

Discuss any special equipment that would be required to do each job.

Suggest numbers of each type of labour and draw a bar chart to demonstrate the economy resulting from your choice.

Discuss the relationship between staff, cost and delivery date.

Discuss a likely overhead cost.

Suggest a policy for preventing idleness.

How much would you charge for switches?

How would you control the progress of the work?

Use a bar chart to show how to allocate work, and how much work to aim at getting if you are to use a constant labour force on the design of a sequence of similar (but different) systems.

Bibliography to Part 4

15(a) E. Buffa, *Operations Management, Problems and Models*, Wiley, New York, 1968.
Further Reading
A. Battersby, *Network Analysis for Planning and Scheduling*, MacMillan, London, 1964.
K. Lockyer, *An Introduction to Critical Path Analysis*, Pitman, London, 1964.

Part 5
MODELS AND TOOLS OF ANALYSIS

CHAPTER 16
Failure, Reliability and Life

Costs cannot be calculated without an understanding of life, reliability and failure and at an extremely early stage of the design these subjects are very powerful tools in the selection or rejection of any proposed solution to the designer's problem.

Chapter 16 is intended to be a simple, short introduction to the subjects and has been included because few introductory texts are available. Bazovsky, 1961, and the ARINC Research Corporations publication, 1964, are both useful texts but both are rather long for an undergraduate to use as course books.

When several possible methods of solving a design problem have been listed, the designer has to choose the best of these. In chapter 10 this was discussed briefly but it is obvious that quite extensive calculations may have to be done to show that one system is better than another or that one proposed solution will not work at all. This type of calculation has been accepted as the everyday work of the engineer and most formal courses of instruction have until recently been almost entirely concerned with some classic tools of analysis such as calculations of strength, calculations of performance (electrical, mechanical, hydraulic, aerodynamic or whatever performance concerns our system) and geometry. More recently, the systems approach and the critical approach to value has led to the employment of men with deep specialist knowledge in fields only briefly mentioned in classic courses of instruction.

One such specialization is Quality and Reliability. Years ago, Quality Control went little further than checking a detail part to ensure that, geometrically, it was to drawing. With more sophisticated systems operating in ever more difficult environments, the costs of failures and unreliability became conspicuous and the sciences of Quality Control and Reliability mushroomed.

FAILURE ANALYSIS

It is always desirable to make a judgment as early as possible. If a design scheme is to be rejected because it is unreliable, then the sooner we reject that scheme, the better (and cheaper). Failure Analysis is a coarse method of judgment which may usually be applied as a prelude to reliability calculations.

Obvious examples of systems in which considerations of failure are important are the safety valve on a steam boiler and the signalling system on a railway. If a safety valve has to fail, we would almost certainly prefer it to fail open, losing steam rather than permitting the boiler to explode. If a railway signalling system has to fail we would probably prefer it to fail so as to stop all the trains rather than to permit them to run about uncontrolled.

Where failures must be considered it is often possible to do so at a very early stage of the design. A failure analysis can frequently reject an unsuitable design at this early stage when very little money has been spent. As an example, consider a system that was required to warn an operator that an atomic pile refuelling machine was not vertical. The first suggestion was that shown in Fig. 16.1.

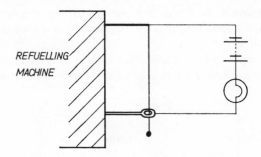

REFUELLING
MACHINE

Fig. 16.1

The bob is suspended by conducting cord and if the refuelling machine is not vertical, the cord touches the conducting ring to complete an electrical circuit which lights the warning lamp. A very few seconds thought leads us to realize that this system is not satisfactory because a failure of the bulb would pass unnoticed, so that it would be possible for an operator to use the machine when it was not vertical and no indication of this fault would be given. An improvement to the design would be the system shown in Fig. 16.2.

Here, departure from the vertical energizes relay A/1. Relay A/1 contains switch A1, which is normally made but which is broken when the relay is energized. Thus the light is on when the machine is vertical and if the light were to go out, a failure would be indicated.

Usually, of course, systems are more complex and a failure analysis correspondingly more difficult. We must establish a formal procedure to ensure that a failure analysis is properly done and we must decide, in advance,

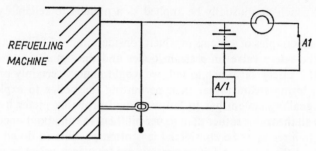

Fig. 16.2

how detailed we wish our failure analysis to be. If we are considering the design at an early stage then we merely wish to demonstrate that our system can be made to work. We may well have to repeat the failure analysis at a later stage to prove to ourselves that our details will also be acceptable. A sensible technique, best demonstrated by example, is shown in Fig. 16.3 and the following table.

Fig. 16.3 is similar to Fig 2.1 which has been used as an example on several occasions but we use here a sketch evolved at an even earlier design stage.

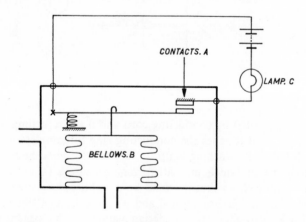

Fig. 16.3

Assume that the system of Fig. 16.3 has been suggested as a means of providing a warning that the pressure in the oil system of a prime mover has fallen below a safe value. Such a system is used on most motor cars to indicate a low oil pressure but may be used in large-scale prime movers in, for example, aeroplanes or power stations. The oil pressure is applied to the outside of the

*bellows B and when acceptably high, causes a contraction of the bellows and
a breaking of the contacts A so that the lamp C goes out.
Some failures are considered in the following table:*

Component	Failure	Result of failure	Indication of failure	Action	Remarks
1.1 Contacts, A	Fails open circuit	Lamp cannot light up	Lamp does not light before starting the engine	Do not start engine until warning system has been checked	Apparent only at start-up because lamp should light before engine starts. Acceptable if engine is frequently started and run for short periods only
1.2 Contacts, A	Fails short circuit	Lamp lights up whether there is a failure of the engine or not	Lamp lights up	Stop engine	Failure is indicated
2 Bellows, B	Fractures	Bellows expand and lamp lights up	Lamp lights up	Stop engine	Failure indicated
3 Lamp, C	Fails·	Lamp will not light up if there is an engine failure	Lamp does not light up before engine is started	Do not start engine until warning system has been checked	Apparent only at start-up. Acceptable if engine is frequently started and run for short periods only

*Note that the analysis is not exhaustive and many more possible failures could and
should be considered. Consider for example a fracture of the case, a fracture of the pipes,
a blockage of the pipes, etc.*

It is worth remarking that, if we assume the system to be the oil pressure warning system of a motor car, then the failures examined are probably acceptable. If, however, we were considering an oil pressure warning system for a very expensive prime mover (say a gas turbine worth £100,000) operating for very long periods (say in a power station) then it is unlikely that 1.1 and 3 would be acceptable since these failures would not be detected except at start-up. These failures if they occurred during running would lead to the possibility of an engine running without a useful oil pressure warning system and hence the possibility of an expensive engine failure.

Note that the table poses the following questions:

What failure are we considering?

What is the result of the failure?

How is the failure indicated?

What should be done if a failure is indicated?

Can we accept the situation?

A failure analysis that does not answer these questions is useless. To examine a failure we must know what damage it does; if a failure is significant we must know that it has occurred; if we know that it has occurred we must be able to do something about it or we might just as well not know.

Suppose we were unable to accept the situation described in the remarks column of our table. Suppose that we wanted to be sure that lack of indication of failure could not mislead us. Then we would clearly have to modify the design. At this stage in the procedure we can afford to do so because our scheme is little more than a free-hand sketch, executed in a matter of minutes, and the analysis itself will have cost only half an hour or so.

In more complex systems, a failure analysis can still be done at a very early stage but will obviously take more time. However, the possible savings are increased with the value of the system. There are many cases in history where unacceptable failures have only been discovered at an advanced stage of design and where the remedy has been a complex one grafted on to what is a fundamentally wrong conception. A failure analysis at the right time can save very large sums of money which might otherwise have to be spent later to rectify a fault discovered in service.

It is sometimes difficult to avoid the possibility of a failure that can pass unnoticed and the mere fact that the failure passes unnoticed must mean that, in the circumstances which prevail, the failure has not been catastrophic. If we really cannot ensure that such a failure is advertised then we will have to consider its effect in combination with any other possible failure. This is the only circumstance in which designers commonly consider the result of two simultaneous failures. The only justification for not normally considering combinations of failures is that life is short, time is valuable and the chance of two failures occurring simultaneously is very much less than the chance of one failure occurring. Once one considers the chance of a failure occurring then one is really embarking on a calculation of reliability, an analysis usually done later in the design process.

There are occasions too when the designer finds it impossible to eliminate possible failures at an acceptable cost. It may be necessary to abandon the project but in many cases it may be desirable to calculate the risk of failure and weigh the cost and probability of such a failure against the cost of avoiding it. In the case of a car, a leaking radiator or loss of oil could cause expensive failures. Guarding against such failures is also expensive, however, and many motorists are prepared to take the chance of a failure, a chance which can be reduced by checking water and oil levels daily.

RELIABILITY

As we have seen, a failure analysis is a go/no-go gauge which may be applied to a proposed solution of our problem at a very early stage of conception. It is however, a comparatively crude criterion of acceptance and if our system is still acceptable after a failure analysis we will need to check its reliability. Unlike the failure analysis, reliability analysis must generally be deferred to an advanced stage of design (a possible exception is the design of a system using all well-understood, existing components, when our design may be little more than a block diagram).

We may wish to predict the reliability of a system for one or two reasons. We may be obsessed by safety considerations and yet be unable to ensure that an unsafe failure is impossible. Under such circumstances we will attempt to design so that the chances of the dangerous failure occurring are so remote that we will be prepared to accept whatever risk is involved. Consider the structure of a passenger-carrying aeroplane. It is possible to point to several parts (e.g. the wing main spar, the connection of the empennage to the fuselage, the fuselage itself) of which the failure would almost certainly cause the aeroplane to crash. Whatever design we adopt, we are unable to ensure that no structural failure will be catastrophic but we can design so that our structure has a very high reliability. Thus, few designers would design an aeroplane such that if the wing broke off there would be no danger but thousands of people every day are prepared to accept the risk that this will happen.

Alternatively, safety may not weigh very heavily with the designer and reliability then becomes entirely a matter of cost. An unreliable motor car might be cheaper to produce than a reliable one but it would be difficult to sell in competition. Further, the costs of repairing the unreliable car would probably mean that the purchaser would eventually pay more for each mile of motoring. The builder of the less reliable car would probably make less profit in the long run than the builder of the more reliable car. Clearly though, some compromise is necessary since the ultimate in reliability would probably be too expensive to sell.

Any one of several cases may offer itself. If our system uses all new parts we may have no practical experience which will enable us to predict the reliability of our system or even of any part within it. In such a case we may build in ignorance and use only formal stress calculations to convince ourselves that what we have designed has a reasonable chance of an acceptable reliability; this we will find out only after our system has been in service for a shorter or longer period. Sometimes however, we may be able to find old parts which have sufficient in common with the parts of our system for us to use 'read across' evidence of reliability to increase our confidence.

At the other extreme, our system may use all components of which a great deal is known of their behaviour in other, earlier systems. This is often true of electronic systems, in which we use transistors, capacitances, resistances, connectors etc., which have already been used in earlier circuits. In this case, knowledge of the reliability of the components in earlier systems enables us to predict the reliability of our new system.

Generally, of course, the truth lies between these two extremes and we find that we have designed a system composed of parts, some having a known reliability record, some which are new but have similarities with parts of known history and some for which we can only guess the reliabilities.

Because we need to know the reliabilities of the components before we can calculate the reliability of a system, it is important that we do all we can to log the service of any system to provide reliability data for all components. Not only does the analysis of such data enable us to predict the reliability of future systems using the same components but it also shows us where redesign is desirable. A system for checking the reliability of parts in service implies a feedback of trouble reports to the designers and the identifying, by nameplates, of all parts in the field (cf. chapters 11 and 12).

To measure reliability implies that it is quantitative and normally reliability is a number derived from the failure rate of the component. Commonly, newly-made designed components fail frequently because we have made mistakes in design or we have not yet learnt to manufacture or use them properly. After an initial 'debugging' period however, we correct our faults and the failure rate of the components usually improves. Finally, when components begin to wear out the failure rate increases. Even when we have been producing a component for long enough to have corrected design faults, a similar failure pattern is apparent. Newly-built components tend to have a high failure rate because of undetected manufacturing errors; components which survive this early period fail for chance reasons only; while eventually, surviving components fail because they wear out.

The above phenomenon may be demonstrated by the curve of Fig. 16.4. We make and supply a batch of N systems at time $t = 0$. At any time t the number of systems still operating is n and we plot n against t.

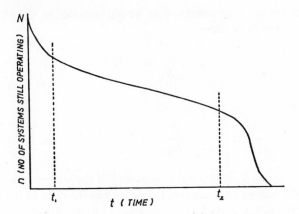

Fig. 16.4 Population vs. time

Initially n falls rapidly but between times t_1 and t_2 failures are due to chance reasons and n falls at a comparatively low rate. After time t_2, n falls rapidly again because systems are now wearing out.

If we consider within the interval (t_1, t_2), the small increment of time $(t, t + \Delta t)$, then at the beginning of this increment we have a population of n systems while at the end of the increment we have a population of $(n - \Delta n)$ operable systems since Δn have failed.

The number of systems which fail will be proportional to the number at the beginning of the increment and also to the length of the increment. We may therefore write, for the period (t_1, t_2) that

$$\Delta n \propto \Delta t \times n$$

or

$$\Delta n = \lambda n. \Delta t \tag{16.1}$$

where λ, the constant of proportionality, is called the failure rate. Note that when we say that λ is a constant we are assuming that failures are for chance reasons. If we were in a time interval where the systems were wearing out, we would have a value of λ which increased with time.

If we take limiting values of Δn and Δt, equation (16.1) may be rewritten

$$\frac{dn}{dt} = -\lambda n \quad \left(\text{note } \frac{\Delta n}{\Delta t} \to -\frac{dn}{dt} \right)$$

and this expression integrates to give

$$\log n = -\lambda t + \text{constant} \tag{16.2}$$

If at the beginning of the interval that we are considering $n = N_B$ and $t = 0$ then equation (16.2) becomes

$$\log \frac{n}{N_B} = -\lambda t$$

or

$$\frac{n}{N_B} = e^{-\lambda t} \tag{16.3}$$

The expression n/N_B is called the reliability of the system. We see that the reliability of a system over a period t is defined as the ratio of the number of systems surviving the period to the original number of systems.

The information in Fig. 16.4 is frequently presented by plotting λ against time and in the case that we have considered, this would yield a curve of the form shown in Fig. 16.5 with a high initial failure rate due to undetected faults of design or manufacture, a constant, relatively low failure rate during the useful life of the system and an eventual high failure rate as the systems wear out.

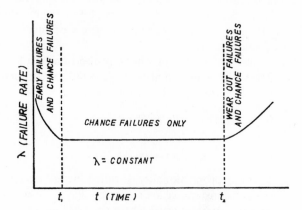

Fig. 16.5 Bath-tub curve

The curve of Fig. 16.5, because of its shape, is frequently known as the 'bath-tub' curve.

Once the initial 'debugging' period is over, we can assume that our system will have an approximately constant failure rate up to the time that parts start wearing out so that we can 'life' our systems. That is, we may tell the customer that the system is good for a life of so many hours, during which he can expect a low, predictable, failure rate from chance causes, but after which the failure rate will be unacceptably high. Although we may choose to guarantee the advertised life of our system, we will, nevertheless, have to expect some failures during that life. Thus we may 'life' a system for a thousand hours so that every one of our systems is taken out of service when it has operated for a thousand hours but every so often, a system will fail before it is taken out of service.

Where guarantees of life are important and failures expensive, the manufacturer may run his systems at the factory for the initial debugging period in order that when the surviving systems are delivered to the customer they will be systems for which the initial high failure rate is past and for which the constant value of λ can be quoted.

In other cases, the customer is allowed to suffer the high failure rate of brand new systems. The cheap car is an example of this. Anyone who has bought a new car knows that he must expect a high rate of component failure during the first few hundred miles and the manufacturer finds it cheaper to use the customer as an inspector and put failures right under guarantee than debug cars before sale.

Chance failures may be expressed in terms of the Mean Time Between Failures (usually abbreviated to M.T.B.F.) in the following way:

Consider a period Δt which we commence with n systems and during which Δn systems fail then

$$\text{M.T.B.F.} = \frac{(n - \Delta n)\,\Delta t + \Delta n \sigma \Delta t}{\Delta n} \qquad (16.4)$$

where σ is a number between 0 and 1. The numerator of this fraction is simply the sum of all the useful lives of the n systems during the period. If all the systems survived, this would be $n\Delta t$ but Δn of the systems have lives less than Δt.

In the limit our expression becomes

$$\text{M.T.B.F.} = \frac{-1}{\dfrac{dn}{dt} \times \dfrac{1}{n}} = +\frac{1}{\lambda} \qquad (16.5)$$

If, then, we know the failure rate (or the mean time between failures) of a system we are able to predict its reliability for a stated period; that is, we are able to state the expected proportion of systems which will survive through the period.

In many cases we build our new system using components which have already been in service in other circumstances. We may therefore have reliability data, from service experience, of the components of our system. By synthesis, we are then able to calculate the reliability of our new system.

In the simplest case our system will consist of n independent components and the system will be considered to have failed if any one of the components should fail. In the context of reliability calculations, we say that the components are in series. If the reliability of the ith component is R_i then the reliability of the whole system is

$$R_{\text{system}} = R_1 \times R_2 \times R_3 \times \ldots \times R_i \times \ldots \times R_n \qquad (16.6)$$

Readers with at least a slight knowledge of probability will recognize that this derives from the fact that the chance of several events occurring together is the product of the chances of the events occurring separately.

Suppose we have a system of two components in series.

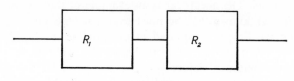

Fig. 16.6

Let the reliability of the first component be R_1 and the reliability of the second be R_2. Since the components are in series they must both survive if the system is to work. Suppose $R_1 = 0\cdot997$ and $R_2 = 0\cdot983$. These figures mean that for the interval being considered, if we had 1000 of each component 997 of the first and 983 of the second would be expected to survive or the first component has 997 chances in a thousand of surviving and the second has 983.

Let us simulate the first component with a thousand-sided die of which 997 sides are labelled S (for success) and 3 sides are labelled F (for fail), then throwing this die will result in S with the same frequency as the survival of the first component.

We can simulate the second component with a thousand-sided die of which 983 sides are labelled S and 27 sides are labelled F.

We can simulate the whole system by throwing both dice. For the system to survive the period, both components must survive and this is simulated if both dice simultaneously show S. The number of possible combinations of dice sides is 1000×1000 and each combination has an equal probability of occurring. The number of combinations of dice sides in which both are S, is 997×983 so that the chance of both dice showing S is $(997 \times 983)/(1000 \times 1000)$. Hence the probability that the whole system will work for the specified period is $0\cdot997 \times 0\cdot983$ or $R_1 \times R_2$.

A simple extension of this idea to n components in series tells us that

$$R_{\text{system}} = R_1 \times R_2 \times \ldots \times R_n \qquad (16.7)$$

or $\qquad R_{\text{system}} = e^{-\lambda_1 t} \times e^{-\lambda_2 t} \times \dots e^{-\lambda_i t} \dots \times e^{-\lambda_n t}$

$$= e^{-(\lambda_1 + \lambda_2 + \dots + \lambda_i + \dots + \lambda_n)t} \qquad (16.8)$$

As a very simple example of components in series, consider an aeroplane with two engines, both of which must operate. If the engines are identical and each has a M.T.B.F. of 10,000 hr then λ for each engine is 0·0001. The reliability of the engine system is therefore

$$R = e^{-(0·0002)} \quad \text{for a one hour flight}$$

$$\simeq 0·9998$$

and $\qquad R = e^{-(0·0002) \times 4}$ for a four hour flight

$$\simeq 0·9992$$

If, having calculated our system reliability, we find that it is too low to be acceptable, we may duplicate parts of low reliability. If, for example, we felt that the bulb in the light which warns of insufficient oil pressure in a motor car engine were of too low a reliability, we might choose to have an oil pressure gauge in addition. The chance that both the devices will fail together is much lower than the chance that the warning light will fail. When we duplicate components in this way we are said to build-in 'redundancy'.

Redundancy may be parallel, as in the example quoted, where either channel is always capable of operating. Symbolically this is drawn as in Fig. 16.7.

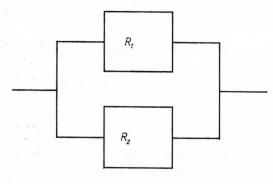

Fig. 16.7

If R is the probability of success ($=$ Reliability) then Q, the probability of failure $= 1 - R$. In our example of the system to warn the motorist of low oil pressure, let us have the warning light and the gauge, in parallel. The system then works if:

the warning light operates and the gauge operates,

the warning light operates but the gauge fails,

the warning light fails but the gauge operates.

Failure results if:

the warning light fails and the gauge fails.

If R_1 is the reliability of the warning light and if R_2 is the reliability of the gauge then the chance of failure of both is

$$Q_1 \times Q_2 = (1 - R_1) \times (1 - R_2) \qquad (16.9)$$

but all other possibilities are counted success so that the chance of success is

$$1 - (Q_1 \times Q_2) = 1 - (1 - R_1) \times (1 - R_2) \qquad (16.10)$$

So that total reliability of two components in parallel is

$$R_{\text{system}} = R_1 + R_2 - (R_1 \times R_2) \qquad (16.11)$$

If, in our earlier example, the aeroplane had been able to fly on one engine then the chance of completing a four-hour journey would have been

$$R = 2e^{(-0.0001) \times 4} - e^{(-0.0001) 2 \times 4}$$

$$= 2e^{(-0.0004)} - e^{(-0.0008)}$$

which differs from 1 only in the 5th decimal figure whereas the reliability of each engine would be

$$R_1 = e^{(-0.0001) 4}$$

$$\simeq 0.9996$$

Note that, in reliability calculations, R frequently differs from one, only in the fifth or sixth decimal place. Tables must therefore be very accurate if they are to be used to give us values of $e^{-\lambda t}$. However, we know that

$$e^{-\lambda t} = 1 - \lambda t + \frac{(\lambda t)^2}{2!} - \frac{(\lambda t)^3}{3!} + \text{etc.} \qquad (16.12)$$

so that when λt is very small, $e^{-\lambda t} = 1 - \lambda t$ with an error which is of the order of $(\lambda t)^2$.

Redundancy may be Standby Redundancy. In this case the redundant component is not normally in the system and a decision-making device is employed to switch in the redundant component when the prime component fails. A simple case, familiar in everyday life, is the braking system of a motor car. The handbrake is a redundant component which may be used if the decision-making device (in this case the driver) decides that the footbrake has failed. Diagrammatically we may represent standby redundance as in Fig. 16.8.

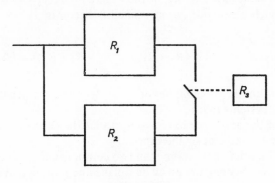

Fig. 16.8

Let the reliability of the Prime Component be R_1
Let the reliability of the Redundant Component be R_2
Let the reliability of the Decision-Making device be R_3
The system will operate successfully if:
    (a) Prime Component and Redundant Component both work and either is in circuit.

$$\text{The probability of this happening is } R_1 \times R_2. \qquad (16.13)$$

or

    (b) The Prime Component works, the Redudant Component does not work but the correct decision is made by the decision-making device.

$$\text{The probability of this happening is } R_1(1-R_2)R_3 \qquad (16.14)$$

or

    (c) The Prime Component fails, the Redundant Component works and the correct decision is made by the decision-making device.

$$\text{The probability of this happening is } R_2(1-R_1)R_3 \qquad (16.15)$$

These three, mutually exclusive events are the only cases that may be considered successful so that the probability of success is the probability that one of these three events occurs, i.e.

$$(R_1 \times R_2) + R_1(1 - R_2)R_3 + R_2(1 - R_1)R_3 \qquad (16.16)$$

The reliability of the system is therefore

$$R_{\text{SYSTEM}} = (R_1 \times R_2) + (R_1 \times R_3) + (R_2 \times R_3) - 2(R_1 \times R_2 \times R_3) \quad (16.17)$$

In some cases the reliability of our decision-making device may differ according to whether the PRIME or the REDUNDANT COMPONENT is in circuit. In such a case our system reliability becomes

$$R_{\text{SYSTEM}} =$$
$$= (R_1 \times R_2) + (R_1 \times R_{32}) + (R_2 \times R_{31}) - (R_1 \times R_2 \times R_{32}) - (R_1 \times R_2 \times R_{31})$$
$$(16.18)$$

where R_{31}, R_{32} are the probabilities of selecting correctly when the prime and the redundant systems fail.

Note that in the above we have assumed that the decision-making device will select one component or the other.

On occasion we may have more than two components in parallel. We may for example have four engines on an aeroplane and require that at least three should work.

Consider that we have n components of reliabilities, R_1, R_2, R_3, ... R_i, ..., R_n. Let us assume that, for success, k of the components must be operating. This means that we achieve success if

n components operate,
$(n-1)$ components operate,
$(n-2)$ components operate
....
....

$(k+1)$ components operate
k components operate

Failure arises if

$k-1$ components operate
$k-2$ components operate
....
....
....
1 component operates
0 components operate.

The chances of n components operating are

$$R_1 \times R_2 \times R_3 \times ... \times R_i \times ... \times R_n \qquad (16.19)$$

The chances of exactly $(n-1)$ components operating are

$$Q_1 \times R_2 \times R_3 \times \ldots \times R_i \times \ldots \times R_n$$
$$+ R_1 \times Q_2 \times R_3 \times \ldots \times R_i \times \ldots \times R_n$$
$$+ R_1 \times R_2 \times Q_3 \times \ldots \times R_i \times \ldots \times R_n$$
$$+ \text{etc.} \tag{16.20}$$

The chances of exactly $(n-2)$ components operating is the sum of all the possible expressions of the form

$$R_1 \times R_2 \times \ldots \times Q_p \times \cdots \times Q_s \times \cdots R_i \times \ldots \times R_n \tag{16.21}$$

i.e. all those products in which there are two Q terms.

And so on.

Clearly the calculation is tedious for the general case although not difficult if we are dealing with only three or four components.

Luckily, the most common case of multiplication is with a number of identical components. The four engines of an aircraft would usually be similar. Three or four pumps in a pumping station would usually be similar. Usually then we are dealing with the calculation of the reliability of a system of n identical components of which k must work.

If the reliability of each component is R then the chances of all working is

$$R \times R \times R \times \ldots \times R \times \ldots \times R = R^n \tag{16.22}$$

The chances of $n-1$ continuing to work are

$$Q \times R \times R \times \ldots \times R \times \ldots \times R$$
$$+ R \times Q \times R \times \ldots \times R \times \ldots \times R$$
$$+ R \times R \times Q \times \ldots \times R \times \ldots \times R$$
$$+ \text{etc.} \qquad = nQR^{n-1} \tag{16.23}$$

The chances of $n-2$ continuing to work will be seen to be

$$= \tfrac{1}{2}n(n-1)\,Q^2 R^{n-2} \text{ etc.} \tag{16.24}$$

In general we see that the chances of any given number of the components failing will be the appropriate term of the expansion:

$$(R = Q)^n = 1 \tag{16.25}$$

i.e. $R^n + nQR^{n-1} + \tfrac{1}{2}n(n-1)\,Q^2 R^{n-2} + \dfrac{1}{3!}\,n(n-1)(n-2)\,Q^3 R^{n-3} + \text{etc.} = 1$

$$\tag{16.26}$$

Example (from the *Boeing Reliability Guide to Vendors*).

Calculate the power plant reliability of a four-engine airplane, if operation of any two engines will sustain flight and the individual probability of each engine operating for the entire mission time is $0 \cdot 9$, assuming independence.

Solution: $n = 4$ $R = 0.9$ $Q = 1 - 0.9 = 0.1$

$$(R+Q)^4 = R^4 + 4R^3Q + 6R^2Q^2 + 4RQ^3 + Q^4 = 1$$

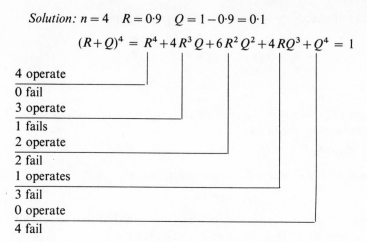

4 operate	
0 fail	
3 operate	
1 fails	
2 operate	
2 fail	
1 operates	
3 fail	
0 operate	
4 fail	

(The sum of the first 3 terms gives the probability that at least 2 engines are operable, hence the probability of mission success).

Probability of success $= R^4 + 4R^3Q + 6R^2Q^2$

$= (0.9)^4 + 4(0.9)^3(0.1) + 6(0.9)^2(0.1)^2$

$= 0.9963$

With a slight knowledge of probability then, and with the ability to predict the reliability of individual components, we can predict the reliability of a system. If the reliability of a system is not adequate then we may modify it by adding redundant components so that its reliability becomes adequate. Adding components usually means adding cost and increasing complication, so that once again the designer is faced with making a compromise and balancing the cost of unreliability against the cost of complexity.

Being able to make reliability calculations at all presupposes knowledge of the failure rates of components and this requires a constant, orderly feedback of information from the field. Part of the work of the Service Engineer must be to feed back complete information about failures in service and the Reliability Engineer will have to catalogue and analyse the information so that as much knowledge as possible of failure rates is always available to the designer.

LIFE

Where information does not already exist it will be desirable to obtain it by means of properly designed tests. Where the life and reliability of a component are not thought to be of extreme importance it may be possible to wait for the customer to complain but where the component is part of a system of which the life and reliability are important, it may be necessary to conduct tests while the system is in service, extending the advertised life of the system as laboratory tests on components permit. Since sometimes the only way of determining the life of a component is to test it to destruction, it can take 1000 hours of laboratory work to demonstrate a life of 1000 hours.

One difficulty in obtaining life or reliability data is that much of what happens has a chance element. If for example, we desire to show that a component is to have a life of 1000 hours it is not enough to subject one such component to simulated operating conditions for 1000 hours without failure. If we were to subject a number of similar components to similar conditions until they all failed, we would in all probability find that each failed at a different life. The first component might last over 1000 hours but the second might fail after less time, the third at a different time again etc.

Let us consider the type of distribution of life that we could expect.

Suppose that we have no reason to believe that λ, the failure rate, is constant. We may suppose that λ is an unknown function of time so that $\lambda = \lambda(t)$.

We would still expect the number of failures in a population to be proportional to the population and to any very short time interval considered thus:

$$\Delta n \text{ (number of failures in population } n)$$

$$= \lambda(t) \times n \times \Delta t \tag{16.27}$$

in the limit, as Δt decreases,

$$\frac{dn}{dt} = -\lambda(t) \times n \tag{16.28}$$

$$\text{or} \int_{N_0}^{N} \frac{dn}{n} = -\int_{0}^{t} \lambda(\tau)\,d\tau \tag{16.29}$$

$$\text{so that } \log \frac{N}{N_0} = -\psi(t)$$

$$\text{where } -\psi(t) = -\int_{0}^{t} \lambda(\tau)\,d\tau \tag{16.30}$$

(since $\lambda(\tau)$ is not a known function we cannot integrate it)

$$\text{or } \frac{N}{N_0} = e^{-\psi(t)} \tag{16.31}$$

or the Reliability, R that a component will survive time $t = e^{-\psi t}$.   (16.32)

Weibull [16(a)] suggests that any piece of hardware may be considered to consist of a number of components in series (he cites as an example, a chain of many links). If we had n components then the Reliability of the system would be the product of the reliabilities of the individual components. If the n components are identical we would have

$$\text{Reliability} = e^{-n\psi(t)}$$ (16.33)

as the probability of a system surviving for time t.

The probability that a system will fail within time t is, therefore

$$Q(t) = 1 - R = 1 - e^{-n\psi(t)}$$ (16.34)

so that we might expect the cumulative distribution function of failures to be of this form.

For $1 - e^{-n\psi(t)}$ to be acceptable as a cumulative distribution function it must be a non-decreasing function of t. There will also be some value of $t (t_u \geqslant 0)$ at which no failures have occurred. A simple function which permits this is

$$Q(t) = 1 - e^{-n\psi(t)} = 1 - e^{\frac{-(t-t_u)^m}{t_0}}$$ (16.35)

Without loss of generality, this may be re-written

$$Q(t) = 1 - e^{-\left(\frac{t-t_u}{\eta}\right)^m}$$ (16.36)

or

$$1 - Q(t) = e^{-\left(\frac{t-t_u}{\eta}\right)^m}$$

so that

$$\ln \ln \frac{1}{1 - Q(t)} = m \times \ln(t - t_u) - m \times \ln\eta$$ (16.37)

(ln denotes natural logarithm).

If (16.36) were really the cumulative distribution of failures with time then, from (16.37), we would expect a plot of $\ln \ln 1/[1 - Q(t)]$ versus $\ln (t - t_u)$ to be a straight line of slope m and with an intercept on the vertical axis of $-m \ln \eta$. Plenty of evidence has been collected to show that the Weibull cumulative probability distribution of (16.36) very often applies and is particularly useful to describe the results of many experiments in which a number of identical articles have been tested to failure.

Let us assume that we have subjected N articles to an endurance test and
have recorded the time at which each failed. If the nth article fails at time t_n
then we will plot $\ln \ln 1/[1-(n/N)]$ versus $\ln (t_n - t_u)$. Usually we assume that
an article is liable to fail from the time we start to use it so that it is not unreasonable to write $t_u = 0$ and attempt to plot $\ln \ln 1/[1-(n/N)]$ versus $\ln t_n$.

Those engineers who deal frequently with reliability data will use Weibull
graph paper (see Fig. 16.9) which has the vertical scale of n/N marked in
divisions that are linearly proportional to $\ln \ln 1/[1-(n/N)]$ and a horizontal
scale of t_n marked in divisions linearly proportional to $\ln t_n$.

If the Weibull distribution is a reasonable approximation to our failure
pattern, we would expect to be able to draw a straight line that is a reasonable
fit to the points that we have plotted.

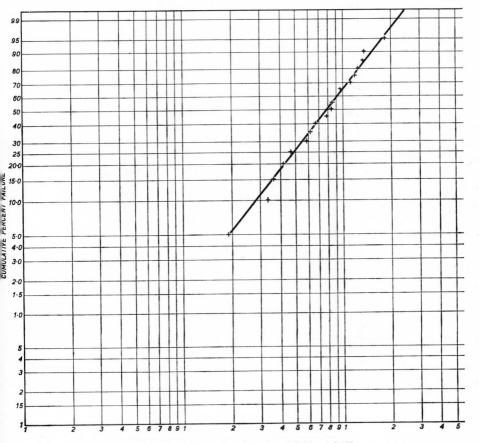

Fig. 16.9 (The data plotted is that of Fig. 16.10)

As an example of such an exercise, consider the following data [taken from 16(b)] of actuations to failure of 20 relays.

1st	relay	failure	occurred at	$1{\cdot}9 \times 10^5$ cycles
2nd	”	”	”	$3{\cdot}34 \times 10^5$ ”
3rd	”	”	”	$3{\cdot}65 \times 10^5$ ”
4th	”	”	”	$4{\cdot}20 \times 10^5$ ”
5th	”	”	”	$4{\cdot}72 \times 10^5$ ”
6th	”	”	”	$5{\cdot}89 \times 10^5$ ”
7th	”	”	”	$6{\cdot}10 \times 10^5$ ”
8th	”	”	”	$6{\cdot}62 \times 10^5$ ”
9th	”	”	”	$7{\cdot}92 \times 10^5$ ”
10th	”	”	”	$8{\cdot}40 \times 10^5$ ”
11th	”	”	”	$8{\cdot}50 \times 10^5$ ”
12th	”	”	”	$9{\cdot}00 \times 10^5$ ”
13th	”	”	”	$9{\cdot}60 \times 10^5$ ”
14th	”	”	”	$11{\cdot}02 \times 10^5$ ”
15th	”	”	”	$11{\cdot}95 \times 10^5$ ”
16th	”	”	”	$12{\cdot}40 \times 10^5$ ”
17th	”	”	”	$13{\cdot}03 \times 10^5$ ”
18th	”	”	”	$13{\cdot}42 \times 10^5$ ”
19th	”	”	”	$18{\cdot}07 \times 10^5$ ”
20th	”	”	”	$20{\cdot}63 \times 10^5$ ”

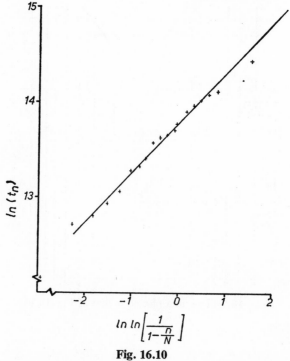

Fig. 16.10

From the above data we have:

n/N	$1/[1-(n/N)]$	$\ln \ln 1/[1-(n/N)]$	t_n	$\ln t_n$
0·05	1·053	−2·956	$1·9 \times 10^5$	12·165
0·10	1·111	−2·254	$3·34 \times 10^5$	12·719
0·15	1·177	−1·814	$3·65 \times 10^5$	12·808
0·20	1·250	−1·501	$4·20 \times 10^5$	12·948
0·25	1·333	—1·249	$4·72 \times 10^5$	13·065
0·30	1·429	−1·030	$5·89 \times 10^5$	13·286
0·35	1·539	−0·842	$6·10 \times 10^5$	13·321
0·40	1·667	−0·672	$6·62 \times 10^5$	13·403
0·45	1·818	−0·515	$7·90 \times 10^5$	13·582
0·50	2·000	−0·367	$8·40 \times 10^5$	13·641
0·55	2·222	−0·226	$8·50 \times 10^5$	13·653
0·60	2·500	−0·088	$9·00 \times 10^5$	13·710
0·65	2·857	0·049	$9·60 \times 10^5$	13·775
0·70	3·333	0·186	$11·02 \times 10^5$	13·913
0·75	4·000	0·327	$11·95 \times 10^5$	13·994
0·80	5·000	0·475	$12·40 \times 10^5$	14·031
0·85	6·667	0·640	$13·03 \times 10^5$	14·081
0·90	10·000	0·834	$13·42 \times 10^5$	14·110
0·95	20·000	1·527	$18·07 \times 10^5$	14·407
1·00	∞	∞	$20·63 \times 10^5$	14·540

These values of $\ln \ln 1/[1-(n/N)]$ and $\ln t_n$ are shown plotted on Fig. 16.10, and we see that it is not unreasonable to fit a straight line to the points plotted.

Usually we can fit an acceptable line to experimental points by eye although we can, if we wish, use a computer program to find the line which is the best fit (i.e. the line for which the sum of the squares of the distances of points from the line is a minimum).

Once the line is drawn, its slope immediately tells us the value of m (by comparison with the line $y = mx + c$).

In the example given above, the line has a slope of $3·5/1·75 = 2$.

We may also determine η, either by the intercept on the vertical axis or by substituting a point on the line in the equation (2).

Considering the point where $\ln \ln 1/[1 - Q(t)] = 0$

and $\ln t_n \quad = 13·75$

we have $\ln \eta = 13·75$

i.e. $\quad \eta = 9·4 \times 10^5$

This means that we believe the failures versus time to have a cumulative distribution function given by

$$Q_t = 1 - e^{-\left(\frac{t}{10^5 \times 9.4}\right)}$$

It is, in fact, possible to extend still further the amount of information that can be obtained from a Weibull plot of failure information.

It can be shown that the mean value of life in the Weibull distribution is

$$= \text{M.T.B.F.}$$

$$= \eta \frac{1}{m} \Gamma\left(\frac{1}{m}\right)$$

where $\Gamma(\alpha)$ is the gamma function $\displaystyle\int_0^\infty e^{-t} t^{\alpha-t} \, dt$ which can be obtained from tables [see e.g. 16(c)].

In the example plotted on Fig. 16.10, we have already discovered that $\eta = 9.4 \times 10^5$ and $m = 2$ so that

$$\begin{aligned}
\text{M.T.B.F.} &= 9.4 \times 10^5 \times \tfrac{1}{2} \, \Gamma(\tfrac{1}{2}) \\
&= 9.4 \times 10^5 \times \tfrac{1}{2} \times 1.77 \\
&= 8.32 \times 10^5 \text{ actuations}
\end{aligned}$$

It is worth noting that if λ is indeed constant then the Weibull distribution becomes

$$Q(t) = 1 - e^{-\lambda t}$$

i.e. $m = 1$

and $\eta = \dfrac{1}{\lambda} = \text{M.T.B.F.}$

By differentiation, we see that the probability distribution function corresponding to (16.36) is

$$f(t) = \left(\frac{m}{\eta}\right)\left(\frac{t}{\eta}\right)^{m-1} e^{-\left(\frac{t}{\eta}\right)^m}, \; t \geqslant 0, \; m > 0, \; \eta > 0 \qquad (16.38)$$

and when $m = 1$, $f(t)$ would be the exponential probability distribution function, as we would expect.

The Weibull probability distribution function is also a fairly close approximation to the normal distribution function for certain values of m and η.

CONFIDENCE LEVELS

If we do not know the M.T.B.F. of a component then we may clearly collect data from the field to tell us the actual lives of a large sample of components or we may arrange laboratory tests to tell us the lives of many components in simulated working conditions. The percentage of failures may be plotted as a Weibull distribution of time (as discussed above) or in any other way that seems fit. Let us assume that we have, in fact, decided to plot the life versus the fraction of population that has failed as a Weibull distribution. That is, we have plotted $\ln \ln 1/[1-(n/N)]$ versus $\ln t_n$ and we have fitted a straight line to the points plotted (as shown on Fig. 16.10). Unless the line passes through all the points (which is unlikely) we cannot say that the component under test obeys, precisely, the law implied by the line drawn. In fact, the most we can say is that about half the units of our sample have better lives than would be suggested by the line and about half have worse lives.

As a further example, consider our pressure switch:

Fig. 16.11 shows the results of a number of tests on metal bellows of the type used in the pressure switch of Fig. 2.1. The data collected from the tests gave the number of cycles to failure of each bellows together with sufficient data to calculate the extreme stress in each cycle. Some trial and error showed that a straight line could be a fit to the experimental data, with reasonable correlation, if log (stress) were plotted against log (cycles to failure). It will be obvious however, that there is still so much scatter that one could not suggest to the customer that the regression line may be used to predict the bellows life.

The customer is not likely to be satisfied if we tell him that half the components we supply will have better lives than the fitted line suggests because this implies that half the components will fail at unpredicted times earlier than the advertised life. Clearly we require to be able to tell the customer that 95% or 99% will have longer lives than some nominal requirement.

When we have obtained our line of best fit to experimental or service life data, we clearly wish to find the band width which we believe will contain a given percentage of all points that could be obtained.

It is not unreasonable to assume that the experimental points will be normally distributed about the line of best fit (regression line). If we have a large number of points then we may calculate the standard error of the points with respect to the line and accept this standard error as a good approximation to the standard deviation of the normal distribution. In the general case, assume that we have a large number, n, of points (x_i, y_i), $(i = 1, 2, \ldots n)$ and that $y = mx + c$ is a line that is seen to be a good fit to those points as in Fig. 16.11.

For every point (x_i, y_i) there will be a point $(x_i, \bar{y}_i)$ where $\bar{y}_i = mx_i + c$.

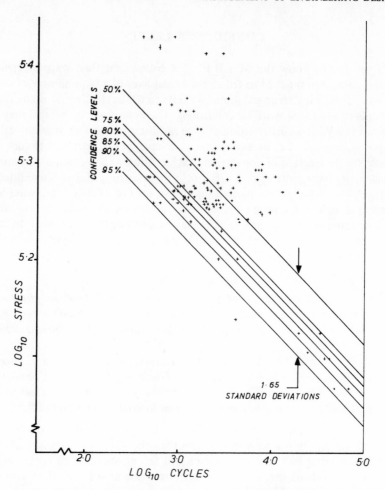

Fig. 16.11 Part of computer-plotted regression lines

The standard error, in terms of the vertical distance of each point from the line is

$$Sy = \sqrt{\frac{\sum_{i=1}^{n} (y_i - \bar{y}_i)^2}{n}}$$

If n is large then Sy may be considered to be a good approximation to the standard deviation of the whole population about the regression line.

If we measure a distance equal to one standard error vertically above and below the regression line then the band so drawn will contain 68% of the population.

Useful values are

± 1·65 standard errors contains 90% of the population
± 1·96 standard errors contains 95% of the population
± 2·58 standard errors contains 99% of the population

Other values may be taken from any table of the areas under the normal curve [see for example, 16(c)].

Fig. 16.11 shows the lines above which are found 50%, 75%, 80%, 85%, 90%, and 95% of the population and these lines define what are called confidence intervals. Consider the 95% confidence line; this is measured 1·65 standard errors below the regression line. ± 1·65 standard errors about the regression line contains 90% of the population but we are solely interested in the fact that only 5% of the population have lives worse than given by the line drawn.

When we have only a small number of experimental points we cannot draw the confidence interval so readily. Common sense tells us that the smaller the number of points obtained by experiment, the wider would our confidence interval have to be. Techniques exist for calculating confidence levels when only a relatively small amount of data is available. Reference may be made to advanced books on probability [see, for example, 16(c), 16(d)] for an explanation of the 't' distribution which enables confidence levels to be estimated when a relatively small sample is tested. For the Weibull distribution, other methods are used since it is usually assumed that both m and η have to be predicted from limited information. In the Weibull case, tables are available for defining confidence levels [see for example 16(e)].

Exercises and Subjects for Discussion

1. Analyse possible failures of your proposed answers to questions 7 and 9, pp. 77.

2. The circuit diagram shows part of a control system. The switch will be moved in response to some error signal and may energize relay A or rest in the position shown or energize relay B. If relay A is energized, contacts $A2$ and $A1$ are made to light lamp $L1$ and energize coil $C1$ (causing the motor to rotate in a clockwise direction). If relay B is energized, contacts $B2$ and $B1$ are made to light lamp $L2$ and energize coil C_2 (causing the motor to rotate in an anti-clockwise direction).

The circuit is only a preliminary suggestion as it is required to devise a system which no single failure prevents from functioning.

Assuming that the relays can fail only in extreme positions and that if relay A fails $A1$ and $A2$ both fail in the same direction, do a failure analysis of the system shown considering only one failure at a time.

Hence by considering a failure of B, suggest a method, using redundant components if necessary, whereby this failure can be tolerated.

3. Discuss how you would ensure that necessary failure analyses are done in any design organization in your charge.

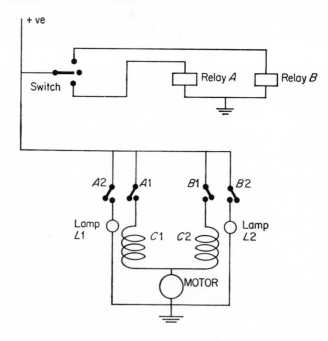

4. (i) Sketch and explain the significance of the so-called 'bath-tub' curve.

(ii) A rocket contains 1000 parts, each of which must work for a successful launch. The M.T.B.F. of each part is assumed to be 10,000 h. If the rocket has to be at readiness for 10 h with all components working, what is the probability of a successful launch?

If three rockets were prepared simultaneously, what is the probability that at least one would be successfully launched? What is the probability that at least two would be successfully launched?

5. For the purpose of a reliability estimate, an aeroplane is assumed to consist of: (i) structure; (ii) power plant; (iii) controls.

The power plant consists of two engines.

Catastrophic structural failure, for chance reasons, is reckoned to occur once every million flying hours.

Catastrophic control failure, for chance reasons, is reckoned to occur once every 100,000 flying hours.

Engine failure, for chance reasons, is reckoned to occur once every 1000 flying hours.

For the first 6 min of every flight, both engines must function or the aeroplane will crash. Thereafter the aeroplane can fly satisfactorily with only one engine operational.

What is the probability that the aeroplane will complete a 5-hour flight?

6. The following data were collected from tests to destruction of 10 steel pipe joints:

1st joint failed at 1000 pressure cycles
2nd joint failed at 2120 pressure cycles
3rd joint failed at 3390 pressure cycles
4th joint failed at 4960 pressure cycles
5th joint failed at 6680 pressure cycles
6th joint failed at 8750 pressure cycles
7th joint failed at 11,250 pressure cycles
8th joint failed at 15,200 pressure cycles
9th joint failed at 22,200 pressure cycles

Although the experiment was continued for a total of 50,000 cycles, the 10th sample had not then failed.

What would be the reliability of one such joint over a period of a year if the pipework in the plant were taken to maximum pressure once a day and allowed to return to zero pressure every night?

7. A radar system in its simplest terms consists of an aerial and a 'black box' containing electrics. For the system to work both the 'black box' and the aerial must work.

The failure rate of the 'black box' is 8·68 random failures per thousand hours.

The failure rate of the aerial is 1 random failure per thousand hours.

The system will not work if two aerials are connected to one 'black box' but it will work if two 'black boxes' are connected to one aerial.

A simply redundant black box is therefore possible but a redundant aerial is only possible with a change-over switch.

Investigate the system shown in the diagram knowing that the switch, if it fails, will fail in the position shown. The switch moves to the left if aerial A_2 fails and to the right if aerial A_1 fails.

The reliability of the switch over a 10 h period is 0·99.

What is the reliability of the system in the diagram for a 10 h period?

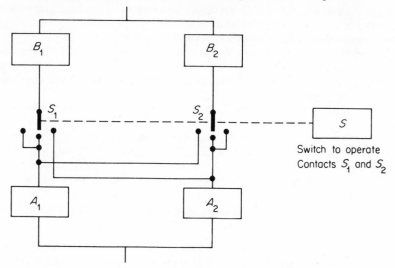

8. (i) Consider any manufactured product with which you are familiar and suggest how there would be financial gain in applying environmental tests to that product. Describe and justify the tests that you would apply.

(ii) A sample of 12 amplifiers was tested to destruction on a vibration table, with the following results:

1st failure at 4·5 hours
2nd failure at 5·7 hours
3rd failure at 6·5 hours
4th failure at 7·5 hours
5th failure at 8·0 hours
6th failure at 9·0 hours
7th failure at 10·2 hours
8th failure at 10·5 hours
9th failure at 11·5 hours
10th failure at 12·0 hours
11th failure at 13·5 hours
12th failure at 20·0 hours

What do you estimate to be the M.T.B.F. of the amplifiers in the vibration environment?

Bibliography

16(a) Weibull, 'A Statistical Distribution Function of Wide Applicability', *J. Appl. Mech.*, **18**, (3) 293 (1951).
16(b) A. Plait, 'The Weibull Distribution', *Industrial Quality Control*, American Society for Quality Control, New York, Nov. 1962.
16(c) E. Kreyszig, *Advanced Engineering Mathematics*, Wiley, New York, 1967.
16(d) D. Aigner, *Principles of Statistical Decision Making*, Collier-MacMillan, New York, 1968.
16(e) M. John and G. Liebermann, *Technometrics*, **8**, No. 1., p. 135. American Statistical Assoc. Richmond, Va., 1966.

Further Reading

I. Bazovsky, *Reliability Theory and Practice*, Prentice Hall, Englewood Cliffs N.J., 1961.
ARINC Research Corporation, *Reliability Engineering*, Prentice Hall, Englewood Cliffs N.J., 1964.

CHAPTER 17
Models and Optimization

Chapter 17 is intended to show that a study of graph theory can lead to a coherent discipline of model building. Some methods of optimization (e.g. dynamic programming) are better introduced and first explained in the context of networks and this relationship between networks and optimization is also mentioned.

It is not intended that the student be provided with useful tools of analysis and no problems have been set.

Flow graphs are used as tools in the analysis of some mechanical and electrical systems and the reader is referred to some such text as Huggins and Entwistle, 1967, for a detailed and useful description of this use.

The use of graphs in the wider, general field of model building is a much more difficult subject. Ford and Fulkerson, 1962, Busacker and Saaty, 1965, and Kaufmann, 1967, are typical of books which place graph theory in a more general context and discuss its relationship with optimization.

If we have a design problem and we propose a system as a solution to that problem, then that system may be cheap or dear to build and try. If the system is cheap to build then we may well build and try it as long as failure to work is not likely to prove expensive. If failure to work properly would be catastrophic or if the system were expensive to build then we would want to convince ourselves that the system was a good one before we placed large sums of money at risk. Many electronic circuit designers have only a vague, quantitative idea of what a circuit will do but it is often cheaper to build the system and try it than to attempt to predict the performance in some other way, say by calculation. We would not, however, build a nuclear power station without first convincing ourselves that it had a very good chance of working. We would not put an aeroplane into service without first convincing ourselves that it had a good chance of success. We would not build a chemical works without first proving the process on a small scale, in the laboratory. Usually, in fact, before investing a large sum of money in the system that has been designed, we invest a small sum in a model in order to convince ourselves that the large investment is likely to give us a good return.

In many cases, of course, the model is a small-scale physical replica of the real thing. The glass tubing, retorts, flasks, etc. in the laboratory will be a small-scale, comparatively cheap system with which we can determine the behaviour of the final system; the small wooden mock-up in the wind tunnel will enable us to predict certain properties of the final aeroplane; the small, experimental, atomic power stations built in the fifties enabled the problems of larger, commercially viable power stations to be understood. In other cases, the physical model may be quite far from the real thing such as, for example, when the designer uses an analogue computer in order to study the behaviour of a non-linear control system. In yet other cases, the model will be an entirely mathematical representation, such as when we use differential equations to model an electrical network.

With a complicated, expensive system, the more that we can predict by paperwork, the cheaper will be our design work, for calculation usually costs less than construction and finding a fault by calculation before construction is almost always cheaper than finding that fault by trial and error on hardware. In other words, it is generally cheaper to make predictions from a mathematical model than from a physical realization. Having proposed a system the designer needs to predict its performance in the assumed environment and with the expected input. He does this in order to demonstrate (or otherwise) that the system meets the specification and sometimes, also, in order to choose optimal values for the parameters of the system.

The designer must also analyse the system to prove that it can be built and operated. Sometimes this analysis enables him to choose optimal values to parameters so that building or operating costs are minimized. Sometimes the designer will analyse the system to demonstrate that it can be built in time using the available resources (as has been discussed in chapters 13, 14 and 15).

With all the mathematical methods for analysing the system, the designer is really using a mathematical model. Most often the initial mathematical model is a graph.

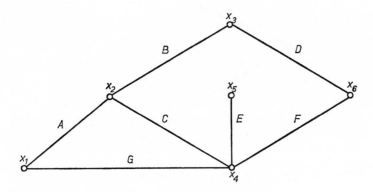

Fig. 17.1

In its most general form the graph is simply a number of vertices connected by a number of edges as in Fig. 17.1.

Here the edges are labelled A, B, C, D, E, F, G and the vertices x_1, x_2, x_3, x_4, x_5, x_6.

Such a graph could be an electrical network in which the edges represented resistance, inductance and capacitance while the vertices represented junctions.

For example we could represent a simple electrical circuit as in Fig. 17.2.

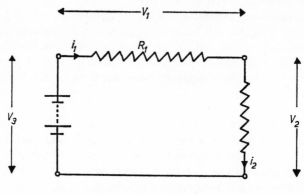

Fig. 17.2

A graph may be used as an aid to calculation, as well as a representation of the system.

For example, Kirchoff's voltage law tells us that in Fig. 17.2 $V_1 + V_2 - V_3 = 0$ and this may be expressed graphically as in Fig. 17.3.

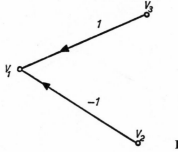

Fig. 17.3

Here the edges are scalar operators and the vertices are the values of voltages while arrows indicate the direction in which the operation is carried out

i.e. $V_1 = V_3 - V_2$

The graph as drawn tells us nothing about V_3 or V_2 in terms of the other parameters. We have arrowed the graph from V_3 because this will be the source voltage of the circuit and we wish to find all else in terms of V_3.

Kirchoff's current law tells us that $i_1 = i_2 = i_3$ and this can be expressed graphically as Fig. 17.4.

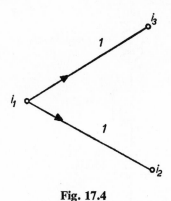

Fig. 17.4

Ohm's law tells us that $i_1 R_1 = V_1$ and this may be expressed by Fig. 17.5.

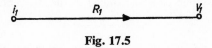

Fig. 17.5

Putting together these ideas we can obtain Fig. 17.6 which with some manipulation will tell us all we need to know about the performance of the circuit.

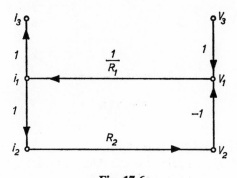

Fig. 17.6

Where we have dynamic problems, we can still use graphs although the operators will not now necessarily be scalar.

If for example we have a viscously damped oscillation from the system of Fig. 17.7

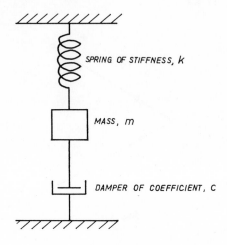

Fig. 17.7

we know that this system can be reduced to the equation of motion

$$m\ddot{x} = -kx - c\dot{x}$$

but the problem could be considered graphically where, if we know how to differentiate, we have Fig. 17.8.

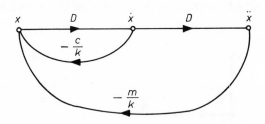

Fig. 17.8

where D denotes differentiation w.r.t. time.

In fact, we can more readily integrate than differentiate on an analogue computer and so we might prefer Fig. 17.9:

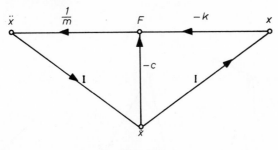

Fig. 17.9

where F denotes the force on the particle and I is the operation of intergrating w.r.t. time.

This last graph could form the basis of an analogue computer solution of the viscously damped spring–mass system because an analogue computer can readily be made to perform all the operations required—integration, addition, multiplication by a scalar.

Calculations involving the solution of many linear algebraic equations such as are often met in structures problems may be studied by means of graphical models.

Algorithms (particularly where digital computers are involved) for the solution of performance problems of a system are often built up graphically as flow diagrams.

When the designer has to study the management of a project he turns to Critical Path Methods, with or without resource limitations, and he uses a network, another manifestation of the graph, already discussed in chapters 13 and 14.

The designer cannot limit his concern to the mechanical performance of a system and the calculation of delivery dates. Problems which also require solution are methods of manufacturing and methods of operating the system. Here again the designer will use graphs.

Fig. 17.10 is a graph showing how to build a main assembly, A. Main assembly A is built of sub-assembly B, sub-assembly C and details P and Q. Sub-assembly B is built of sub-assembly D and details I and J, etc. Such a graph can supply information to the fitter, to the buyer (if numbers of details are put on the edges), to the planner, to the man designing the shop layout, to the programmer scheduling machines etc. Some schedules of drawings incorporate such a graph in their lay-outs.

Similar graphs can show fault-finding procedures, operating procedures, and many other procedures concerned with the man/machine interface of the system. The designer then, will use graphs to model ergonomic problems, to satisfy himself that his system can be built, can be operated, and that neither process will be unnecessarily complicated.

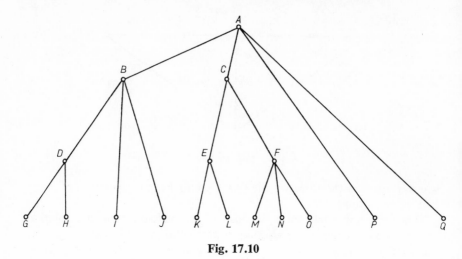

Fig. 17.10

Most good designers think in terms of models and since graphs are often the bases of such models, there is some point in having at least a slight knowledge of graphs. Thinking in terms of graphs gives a logical direction to thoughts. Not only mathematical methods but the logical development of a pilot plant or the computer modelling of a complex manufacturing process may both be manifestations of graphs.

Models are obviously useful for predicting performance but they may also be keys in design optimization. We have already seen that Critical Path Methods are a means of calculating, from a graph, the shortest time in which a project may be completed. Many other optimization techniques may be modelled as longest or shortest routes through networks.

Consider the graph of Fig. 17.11.

This could be a map of the possible roads from town A to town Q, with each number representing the length of a road.*

It could be a diagram of telephone exchanges with the numbers representing the probability of the lines being engaged.*

It could be the critical path network of a project with each number representing the time of an activity.

It could be a decision tree with the numbers representing the pay-offs of certain decisions.

In such cases the designer will want to find the shortest path from A to Q (if the diagram were a map) or the longest path (if the diagram were a critical path network).

---

* In the following discussion of optimizing methods, directed edges have been assumed and this does limit the extent to which these situations may be modelled by the graph of Fig. 17.11.

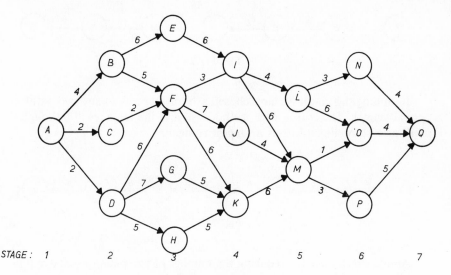

STAGE: 1   2   3   4   5   6   7

Fig. 17.11

A method for finding the longest path has already been discussed in chapter 13 (Critical Path Methods) but other methods are available.

The crudest and most obvious way is by complete enumeration of the possible paths. Here we find the length of every possible path from A to Q and we select the longest (or shortest). This is clearly a foolproof procedure but it is nevertheless rarely an acceptable method because of the number of computations involved. In Fig. 17.11 there are 24 paths from A to Q but this is an artificially simple problem. Consider a problem in which we could, at each stage, choose one of ten alternatives. After only ten stages we would have 10^{10} possible configurations to compute. Clearly any real problem would involve so many computations that we could not consider solving it by complete enumeration, even with the aid of a computer.

Most methods of optimization are judged by the number of computations involved and some methods set out simply to reduce the number of computations that complete enumeration would involve.

One technique, 'branch and bound', considers only the higher (or lower) bound of route lengths between any two vertices. Consider the route LNQ. This is clearly a shorter route from L to Q than LOQ and so, if we are looking for the longest path from A to Q, the path LNQ may be eliminated from any further consideration. Similarly, MPQ must be preferred to MOQ so that MO may be eliminated from further consideration. Proceeding in this way, to eliminate edge after edge, we will finally have only those edges which form part of the longest route (Fig. 17.12).

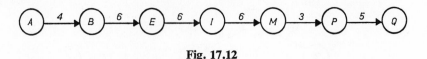

Fig. 17.12

In finding this result we have chosen arbitrarily between any two paths of equal lengths. The critical path method will give the same result although it will also show Fig. 17.13 as a longest route.

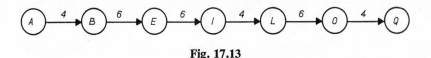

Fig. 17.13

Another technique for reducing the number of computations is 'dynamic programming'. This procedure makes use of the principle that:
'An optimal policy has the property that, whatever the initial state and initial decision are, the remaining decisions must constitute an optimal policy with regard to the state resulting from the first decision.'

If we apply this principle to the diagram of Fig. 17.11 we could say that if we decide to go from A to B as the first stage of the longest path from A to Q, then the remainder of the path, from B to Q must be the longest path B to Q. Although this appears to be a statement of the obvious, the principle is nevertheless powerful in reducing the number of computations that have to be made in determining a sequence of optimal decisions.

Considering the decisions, stage by stage, dynamic programming leads us first to calculate the longest paths to B, C and D. For the second stage decision, we need to consider only the longest paths in stage one to calculate the longest paths to E, F, G and H. These longest paths are used to calculate the longest paths to I, J and K and so on. It is immediately obvious that the number of computations is much less than with complete enumeration because only the longest paths in stage one are considered at stage two whereas complete enumeration would consider all possible stage one paths at stage two.

In the example discussed, dynamic programming is a very similar process to the critical path method. The parallel would be less obvious in a problem that was not first described as a network and generally dynamic programming yields a recurrence relationship between the optimal decision at one stage and that at the next.

Yet another method of finding the longest or shortest path is 'linear programming'. Let T_Q be the length of the longest path from A to Q; T_N the length of the longest path from A to N and so on. We know that Q must be reached via N, O or P so that

$$T_Q \geqslant T_N + 4$$
$$T_Q \geqslant T_O + 4$$
$$T_Q \geqslant T_P + 5$$

with similar inequalities for every other vertex. We could solve the inequalities for T_Q but we would find an infinity of solutions since (as we know from other methods) any value of T_Q that is not less than 30 will serve. Clearly, however, we require the least of these feasible solutions so that our problem can be stated as:

Minimize T_Q

Subject to $T_Q - T_N \geqslant 4$
$$T_Q - T_O \geqslant 4$$
$$\cdot \ \cdot \ \cdot \ \cdot \ \cdot \ \cdot$$
$$\cdot \ \cdot \ \cdot \ \cdot \ \cdot \ \cdot$$
$$T_M - T_I \geqslant 6$$
$$\cdot \ \cdot \ \cdot \ \cdot \ \cdot \ \cdot$$

etc.

The slightly more general problem of optimizing a linear expression in a number of variables, subject to a number of linear inequalities in those variables, is the linear programming problem discussed at greater length in chapter 19.

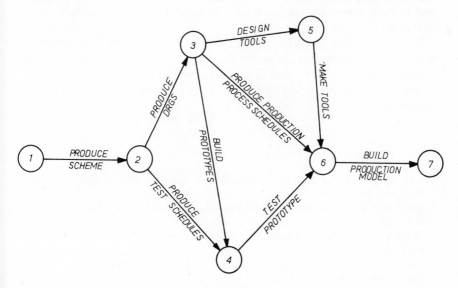

Fig. 17.14

When graphs are used to model complex situations they normally require to be manipulated by computers. The resource-limited management situation described in chapter 14 is typical of such models. Consider the simplified design process modelled in Fig. 17.14.

In most commercial situations, the resources (in this case labour of various skills) will be expected to deal with many jobs, for even if the organization is concerned with the design of only one major system, a large number of sub-systems must be designed simultaneously. We will have then the situation in which each edge of the graph represents a process capable of dealing with similar operations in different systems, although it is likely that only one operation can be carried out at a time. An organization designing an aeroplane will have a structural testing department but the capacity of that department will be limited so that some structural test jobs will have to wait their turn in a queue while others are being worked on.

What we have then in such a model is a graph in which each edge represents an operation and a capacity for work while each vertex represents a list of part-finished jobs waiting to be worked on. The list of jobs at each vertex may, in simple situations, be regarded as a vector of many dimensions, where each component represents a job. In most real cases, however, a vector is inadequate to carry the necessary information. Since each job is different from the others it is usually necessary to carry much more information through the network than a mere name (times of operations, necessary resources, variations in route, the possibilities of operation splitting or continuous feeding are all pieces of information that may have to be carried) so that at each vertex we may well have to carry a many-dimensional array instead of a simple vector.

Rules for selecting jobs from each queue and controls for each operation must also be written into the model so that it becomes a complex means of manipulating large, structured lists of information. The only practical way of dealing with such a model in any but the most simple of situations is with a computer.

Further Reading

W. Huggins and D. Entwistle, *Introductory Systems and Design*, Blaisdell, Waltham, Mass., 1967.

L. Ford and D. Fulkerson, *Flows in Networks*, Princeton Univ. Press, Princeton. N.J., 1962.

R. Busacker and T. Saaty, *Finite Graphs and Networks*, McGraw-Hill, New York, 1965.

A. Kaufmann, *Graphs, Dynamic Programming and Finite Games*, Academic Press, New York and London, 1967.

CHAPTER 18
Optimization and Design

Computer-aided design is a fashionable study and one important aspect of it is optimal design.

Chapter 18 is a review of some of the simpler methods of optimization that are used in design. It will be seen that all the examples considered are trivial and, as it is not expected that the student will acquire directly usable skills, no problems have been set. In any case, optimization is normally applied to design problems where the topology is already known and it is therefore usually restricted to subjects requiring a high degree of local knowledge (as, for example, in the design of the pipe flanges to a British Standard or ship frames for a particular range of tankers).

In our earlier discussions (particularly Part I) we have seen the designer's objective as the fulfilment of a customer's need while various resources and environmental conditions either help or hinder him. The problem may, however, be looked at from a different point of view.

In creating a system, the designer will be making drawings which describe its geometry, schedules which describe its manufacture and schedules which describe its operation and maintenance. In doing this he will be putting numerical values to a large number of parameters. Detail drawings obviously put numerical values to linear dimensions but fuels, life, operating times and very many other parameters must also be given numerical values. The designer's job may be considered as listing all the relevant parameters and giving them the necessary numerical values.

If we call the parameters $x_i (i = 1, 2, ..., n)$ then when all are specified, the performance of the system in any given environment is determined. In the simple case of a damped spring/mass system, the designer will give values to the mass, the spring stiffness and the damping coefficient and the performance of the system in any given vibration environment will be determined.

Similarly, the consumption of any resource is determined when the numerical values of all parameters are given. Given enough data to specify a motor car, its fuel consumption is determined; given enough data to describe a washing machine, the man hours required in its manufacture will be determined; given enough data to describe an aeroplane, its required runway length is determined.

Further, when all the parameters of a system are given numerical values then the cost of building and operating the system is determined. Given enough data to specify a motor car or a washing machine or an aeroplane then their manufacturing costs and running costs are determined.

In theory we can say that the cost of a system, $C = C(x_i)$, meaning that it is a function of the parameters x_i.

We may also say that the required amount of any resource is a function of the parameters, x_i, and must not exceed the amount of the resource available so that

$$B_j(x_i) \leqslant b_j$$

where b_j is the amount of the jth resource available and $B_j(x_i)$ is the amount required.

Further, if the kth required output of the system is p_k and the acceptable limits are $+\delta_k$, $-\delta_k$, we have

$$p_k + \delta_k \geqslant P_k(x_i) \geqslant p_k - \delta_k$$

Where $P_k(x_i)$ is the performance achieved with the parameters x_i.

The designer wishes to provide the required performance at least cost and so his problem may be written:

find the values x_i, $i = 1, 2, ..., n$,

which minimize $C = C(x_i)$ (18.1)

subject to $B_j(x_i) \leqslant b_j, j = 1, 2, ..., m$ (18.2)

and $p_k + \delta_k$ $\geqslant$ $P_k(x_i) \geqslant p_k - \delta_k, k = 1, 2, ..., q$ (18.3)

(The designer may, in fact, be seeking maximum profit rather than minimum cost but a similar argument would apply.)

In this language, the designer's task is an optimization problem in which the objective function is the cost function and the required performance is merely stated as a constraint with no more significance than the constraints imposed by limited resources. Although the question of environment has not been discussed, the argument could be extended to include environment by postulating a value of p, b and δ for each environment.

The statement contained in the expressions (18.1), (18.2) and (18.3) does of course, oversimplify the designer's problem because it omits the creative part of the work, which is the creation of the functions P. Complications also arise because there is usually more than one way of solving a problem so that several sets of P, B, C may be available.

If the designer's problem can be reduced to the optimization of a mathematical function subject to given constraints, then some classical methods are available.

The Unconstrained Problem [18(a)]

If no constraints existed then necessary conditions for $C(x_i)$ to have an extreme value at $(x_i) = (\bar{x}_i)$ are that

$$\frac{\partial C(x_i)}{\partial x_i}\bigg|_{(\bar{x}_i)} = \frac{\partial C(x_i)}{\partial x_2}\bigg|_{(\bar{x}_i)} = \frac{\partial C(x_i)}{\partial x_j}\bigg|_{(\bar{x}_i)} = \frac{\partial C(x_i)}{\partial x_n}\bigg|_{(\bar{x}_i)} = 0 \qquad (18.4)$$

where $(\bar{x}_i)$ implies that the derivatives are evaluated at $x_i = \bar{x}_i$ and it is assumed that $C(x_i)$ is differentiable at $x_i = \bar{x}_i$.

Example (the following example is trivial but it will be seen that as constraints are introduced it becomes progressively more difficult).

A manufacturer wishes to construct a tin can such that the area of tinplate used, $A = 2\pi r^2 + 2\pi rl = A_0$ (r is radius of tin and l its length).

Find r and l which maximize the volume of the tin can.

Although the problem apparently involves a constraint, this may be eliminated by finding l in terms of r, thus:

$$l = \frac{A_0}{2\pi r} - r$$

The volume

$$V = \pi r^2 l$$

$$\therefore V = \pi r^2 \left(\frac{A_0}{2\pi r} - r\right)$$

for an extreme

$$\frac{\mathrm{d}V}{\mathrm{d}r} = \frac{A_0}{2} - 3\pi r^2 = 0$$

i.e.

$$r = \sqrt{\left(\frac{A_0}{6\pi}\right)}$$

also $\mathrm{d}^2 V/\mathrm{d}r^2 = -6\pi r$, which is negative for positive values of r, so that the value of r that we have found corresponds to a maximum value of V. We can clearly go on to find the maximum value of V and the corresponding value of l.

As a further example, where two independent variables are involved, consider the problem of manufacturing batch size. In a simple case of two commodities, it can be shown that

$$C(q_1, q_2) = \frac{c_1 q_1}{2} + \frac{c_2 q_2}{2} + \frac{d_1 r_1}{q_1} + \frac{d_2 r_2}{q_2}$$

where $C(q_1, q_2)$ is the cost of the policy

q_1 is the batch size of commodity 1
q_2 is the batch size of commodity 2
c_1 is the cost/period of storing commodity 1
c_2 is the cost/period of storing commodity 2
r_1 is the period requirement of commodity 1
r_2 is the period requirement of commodity 2
d_1 is the set up cost of making commodity 1
d_2 is the set up cost of making commodity 2

To find values of q_1, q_2 which minimize $C(q_1, q_2)$ we find

$$\frac{\partial C(q_1, q_2)}{\partial q_1} = \frac{c_1}{2} - \frac{d_1 r_1}{q_1^2} = 0 \text{ at an extreme}$$

and

$$\frac{\partial C(q_1, q_2)}{\partial q_2} = \frac{c_2}{2} - \frac{d_2 r_2}{q_2^2} = 0 \text{ at an extreme}$$

so that optimum batch sizes are

$$\bar{q}_1 = \sqrt{\left(\frac{2d_1 r_1}{c_1}\right)}$$

$$\bar{q}_2 = \sqrt{\left(\frac{2d_2 r_2}{c_2}\right)}$$

Constraints Are Equalities [18(a)]

In the slightly less simple case in which all the constraints are equalities we have the method of Lagrange's undetermined multipliers.
Our problem may be rewritten:

Minimize $C(x_i)$ (18.5)

Subject to $G_j(x_i) = 0, \quad j = 1, 2, ..., m.$ (18.6)

We can construct the function

$$F(x_i, \lambda_j) = C(x_i) + \sum_{j=1}^{m} \lambda_j G_j(x_i) \tag{18.7}$$

and necessary conditions for an extreme value of $C(x_i)$ at $(\bar{x}_i)$ are that

$$\left.\frac{\partial F}{\partial x_1}\right|_{(\bar{x}_i)} = \left.\frac{\partial F}{\partial x_2}\right|_{(\bar{x}_i)} = ... = \left.\frac{\partial F}{\partial x_i}\right|_{(\bar{x}_i)} = ... = \left.\frac{\partial F}{\partial x_n}\right|_{(\bar{x}_i)} = 0 \quad (18.8)$$

$$\text{and } \left.\frac{\partial F}{\partial \lambda_1}\right|_{(\bar{x}_i)} = \left.\frac{\partial F}{\partial \lambda_2}\right|_{(\bar{x}_i)} = ... = \left.\frac{\partial F}{\partial \lambda_j}\right|_{(\bar{x}_i)} = ... = \left.\frac{\partial F}{\partial \lambda_m}\right|_{(\bar{x}_i)} = 0 \quad (18.9)$$

assuming C and G to be differentiable at $x_i = \bar{x}_i$.

Example. Let us reconsider the problem of the tin can, using the method of the Lagrange multiplier. Our problem is:

$$\text{Maximize } V = \pi r^2 l$$

$$\text{Subject to } 0 = A_0 - 2\pi r^2 - 2\pi r l$$

$$= G(r, l)$$

We construct the function

$$F(r, l) = V(r, l) + \lambda G(r, l)$$

then

$$\frac{\partial F}{\partial r} = 2\pi r l + \lambda(-4\pi r - 2\pi l)$$

$$\frac{\partial F}{\partial l} = \pi r^2 + \lambda(-2\pi r)$$

and

$$\frac{\partial F}{\partial \lambda} = A_0 - 2\pi r^2 - 2\pi r l$$

For an extreme value of F

$$\frac{\partial F}{\partial r} = \frac{\partial F}{\partial l} = \frac{\partial F}{\partial \lambda} = 0$$

so that

$$2\pi r l = \lambda(4\pi r + 2\pi l)$$

$$\pi r^2 = \lambda(2\pi r)$$

$$A_0 = 2\pi r^2 + 2\pi r l$$

giving

$$\left.\begin{array}{ll} \lambda & = \dfrac{r}{2} \\[2ex] l & = 2r \\[2ex] r & = \sqrt{\dfrac{A_0}{\pi}} \end{array}\right\} \text{ at an extreme value of } V$$

Constraints Are Inequalities [18(b)]

The problem is more difficult when the constraints are inequalities. One method is that of Kuhn and Tucker in which we first rewrite the problem so that all constraints are of the 'less than' form (those of the 'greater than' form may be converted by multiplying both sides of the inequality by -1).

Our problem may be rewritten:

$$\text{Minimize } C(x_i) \tag{18.10}$$

$$\text{Subject to } G_j(x_i) \leqslant g_j, j = 1, 2 ..., m. \tag{18.11}$$

The problem is first restated by adding slack, unknown variables so that the inequalities may be rewritten as equations thus,

$$G'_j(x_i, u_j) = G_j(x_i) - g_j + u_j^2 = 0 \qquad (18.12)$$

We construct the function:

$$F(x_i, \lambda_j, u_j) = C(x_i) + \sum_{j=1}^{m} \lambda_j G'_j(x_i, u_j) \qquad (18.13)$$

and the necessary conditions for an extreme of $C(x_i)$ at $x_i = \bar{x}_i$ are:

$$\left. \frac{\partial F}{\partial x_i} \right|_{(\bar{x}_i)} = 0 \qquad i = 1, 2, ..., n. \qquad (18.14)$$

$$\lambda_j \{ G_j(\bar{x}_i) - g_j \} = 0 \qquad j = 1, 2, ..., m \qquad (18.15)$$

$$G_j(\bar{x}_i) \leqslant g_j \qquad j = 1, 2, ..., m \qquad (18.16)$$

while, if the extreme is a maximum

$$\lambda_j \leqslant 0 \qquad j = 1, 2, ..., m \qquad (18.17)$$

or if the extreme is a minimum

$$\lambda_j \geqslant 0$$

Example. If in our tin can example we introduce a further inequality constraint that the height must not exceed threequarters of the diameter and also say that the amount of material to be used must not exceed A_0 we may rewrite the problem:

$$\text{Maximize } V = \pi r^2 l \qquad (i)$$

$$\text{Subject to } A = 2\pi r^2 + 2\pi r l \leqslant A_0 \qquad (ii)$$

$$l \leqslant \tfrac{3}{2} r \qquad (iii)$$

(ii) and (iii) must be rewritten as equalities thus:

$$0 = 2\pi r^2 + 2\pi r l - A_0 + u_1^2 \qquad (iv)$$

and

$$0 = l - \tfrac{3}{2} r + u_2^2 \qquad (v)$$

$$F(l, r, \lambda_1, \lambda_2, u_1, u_2) = \pi r^2 l + \lambda_1 (2\pi r^2 + 2\pi r l - A_0 - u_1^2) +$$

$$+ \lambda_2 (l - \tfrac{3}{2} r + u_2^2) \qquad (vi)$$

and for a maximum we have the Kuhn–Tucker conditions:

$$\frac{\partial F}{\partial l} = 0 = \pi r^2 + 2\pi r \lambda_1 + \lambda_2 \tag{vii}$$

$$\frac{\partial F}{\partial r} = 0 = 2\pi r l + \lambda_1 (4\pi r + 2\pi l) - \tfrac{3}{2}\lambda_2 \tag{viii}$$

$$0 = \lambda_1 (2\pi r^2 + 2\pi r l - A_0) \tag{ix}$$

$$0 = \lambda_2 (l - \tfrac{3}{2}r) \tag{x}$$

$$2\pi r^2 + 2\pi r l \leqslant A_0 \tag{xi}$$

$$l \leqslant \frac{3r}{2} \tag{xii}$$

$$\lambda_1 \leqslant 0 \tag{xiii}$$

$$\lambda_2 \leqslant 0 \tag{xiv}$$

We can only proceed with the solution if we give a numerical value to A_0.

Assume $A_0 = 10$ units of area. (xv)

Now suppose $\lambda_1 = \lambda_2 = 0$ (xvi)

Then from (vii) $0 = \pi r^2$ (which is clearly absurd for a maximum value of V).

Suppose then $\lambda_1 = 0$, $\lambda_2 \neq 0$ (xvii)

then from (vii) $0 = \pi r^2 + \lambda_2$ (xviii)

and from (viii) $0 = 2\pi r l - \tfrac{3}{2}\lambda_2$ (xix)

and from (x) $0 = l - \tfrac{3}{2}r$ (xx)

but solving (xviii) and (xix) gives

$l = -\tfrac{3}{4}r$ which is not only absurd but conflicts with (xx).

Suppose then $\lambda_1 \neq 0$, $\lambda_2 = 0$

then from (vii) $0 = \pi r^2 + 2\pi r \lambda_1$ (xxi)

from (viii) $0 = 2\pi r l + \lambda_1 (4\pi r + 2\pi l)$ (xxii)

and from (ix) $0 = 2\pi r^2 l + 2\pi r l - 10$ (xxiii)

solving (xxi), (xxii) and (xxiii) we obtain:

$l = -\dfrac{2r}{3}$, which is absurd.

Suppose then $\lambda_1 \neq 0$, $\lambda_2 \neq 0$

then from (ix) $10 = 2\pi r^2 + 2\pi r l$ (xxiv)

and from (x) $l = \dfrac{3r}{2}$ (xxv)

so that $r = \sqrt{\left(\dfrac{2}{\pi}\right)}$ (xxvi)

and $l = \dfrac{3}{2}\sqrt{\left(\dfrac{2}{\pi}\right)}$ (xxvii)

from (vii) and (viii) we also find that λ_1 and λ_2 are both negative so that our values of r and l in equations (xxvi) and (xxvii) correspond to a maximum value of V. The solution to this problem is shown graphically in Fig. 18.1.

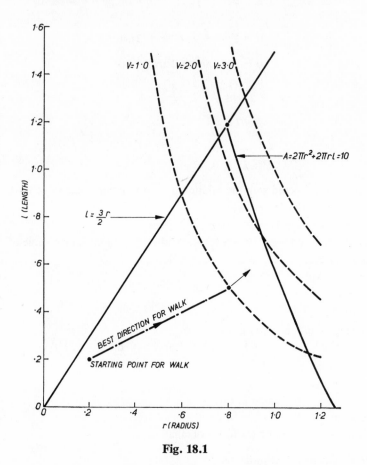

Fig. 18.1

Note that all the above methods assumed $C(x_i)$ and $G(x_i)$ to be differentiable at $x_i = \bar{x}_i$.

In the particular case where $C(x_i)$ $P(x_i)$ and $B(x_i)$ are all linear expressions, the problem is the Linear Programming problem discussed in chapter 19. Linear programming requires special methods because differentiation cannot be used to find the extreme value of a linear function.

Hill Climbing [18(b)]

The most general way of finding an extreme value of the function $C(x_i)$ is a numerical, 'hill climbing' method. We regard $C(x_i)$ as defining the surface of a hill in n dimensional space and having arbitrarily chosen a starting point we 'walk' downhill until there is no lower point (or uphill if we are trying to maximize).

The method may be described most easily by example and where we have only two variables, x_1 and x_2 then

$$C = C(x_1, x_2)$$

represents a family of curves with a curve being defined for each value of C. Each value of C gives, in fact, a contour line on the hill defined by $C(x_1, x_2)$.

Consider the function

$$C(x_1, x_2) = x_1^2 + 2x_2^2 = C$$

We see immediately that the minimum value of C is zero since x_1^2 and x_2^2 cannot be negative but can simultaneously be zero.

$C(x_1, x_2)$ is also seen to be a family of ellipses with an ellipse defined for each value of C as in Fig. 18.2.

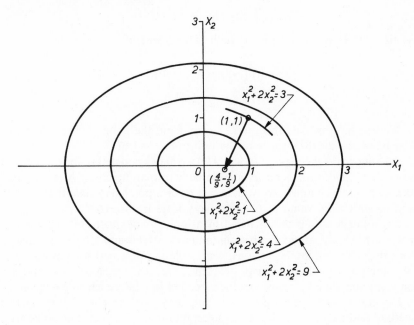

Fig. 18.2

If we choose to start our walk at the point $(1, 1)$ (and assuming that we have no knowledge of the terrain) we could reasonably decide to walk downhill in the direction of the steepest slope. We would set off then, in the direction given by $-\nabla C$, a vector with components $-\partial C/\partial x_1$, $-\partial C/\partial x_2$ (by definition ∇C, called grad C, is the direction of the greatest rate of increase of C).

Since
$$C = x_1^2 + 2x_2^2$$

$$\frac{\partial C}{\partial x_1} = 2x_1 \text{ and } \frac{\partial C}{\partial x_2} = 4x_2$$

so that we would choose to walk in a direction whose components are $-2x_1$ and $-4x_2$, which, evaluated at our starting point, are -2 and -4.

We do not know how far to go but we would clearly reach a point with the co-ordinates

$$(1-2t) \text{ and } (1-4t)$$

where the value of t depends on the distance travelled. If we choose to travel as long as we are going downhill, we can find the lowest point on the path. We know that, on our path

$$C(x_1, x_2) = (1-2t)^2 + 2(1-4t)^2$$

so that the lowest point on the path will be where

$$\frac{dC}{dt} = 2(1 - 2t)(-2) + 4(1 - 4t)(-4) = 0$$

or where $t = 20/72 = 5/18$.

We will have walked then, in one step, from the point $(1, 1)$ to the point $(1-10/18, 1-20/18)$, that is to the point $(4/9, -1/9)$.

Starting again at the point $(4/9, -1/9)$ we may again determine the steepest slope and again walk as far as is reasonable in that direction. After two further steps in this manner, we will, in fact reach a point very close to the origin and at which the value of C differs very little from zero.

The more general problem, in n dimensions, can be treated in the same way and, surprisingly, an acceptable minimum is often found after only three or four steps. The method is tedious, nevertheless, and is best used as the basis of a computer method. In large problems, too, it is necessary to economize on computer time by modifying the direction in which each step is taken.

Difficulties arise because we will find only one local minimum by this method and may fail to identify a global minimum. Further difficulties arise when there are constraints since at each step it is necessary to check that no constraint is broken. Where a desirable length of step would take us beyond a constraint it is necessary to reduce it.

If when walking in the most desirable direction we reach a constraint, it may be reasonable to take the next step within that constraint and the best direction will be the projection of $-\nabla C$ on the constraint but unless the constraint is linear, this may present considerable difficulties.

One method of dealing with constraints introduces a further approximation. Suppose we wish to

$$\text{Minimize } C(x_i) \qquad\qquad (18.18)$$

$$\text{Subject to } G(x_i) \leqslant g, \qquad\qquad (18.19)$$

we may construct the function

$$F_1(x_i) = C(x_i) - \frac{\rho_1}{G(x_i)-g} \qquad\qquad (18.20)$$

where ρ_1 has some small, arbitrary, fixed value. The minimum of $F_1(x_i)$ will provide the starting point of a new walk to minimize

$$F_2(x_i) = C(x_i) - \frac{\rho_2}{G(x_i)-g} \qquad\qquad (18.21)$$

with ρ_2 assuming a smaller value than ρ_1.

The philosophy of this method is apparent when we consider the function $F(x_i)$. When values of x_i are chosen to be distant from the constraint, the small value of ρ ensures that $F(x_i)$ is an approximation to $C(x_i)$. When x_i is chosen to be near the constraint, however, the term $-\rho(G(x_i)-g)$ has a very high value and will prevent any step in a direction tending to break the constraint.

Example. Consider again the problem of the tin can which has already been solved by the Kuhn–Tucker method. For simplicity, we will consider only one constraint although the reader should see how all the constraints could be allowed for.

We will start from a point within the feasible region (such a point is not always easy to find but we will ignore this problem) and arbitrarily we will choose the point

$$r = 0.2, \; l = 0.2.$$

If the problem were unconstrained, we would walk in the direction with components

$$\frac{\partial V}{\partial r}, \frac{\partial V}{\partial l}$$

$$\frac{\partial V}{\partial r} = 2\pi r l = 0.08 \, (\pi)$$

$$\frac{\partial V}{\partial l} = \pi r^2 = 0.04 \, (\pi) \text{ at the point } (0.2, 0.2)$$

Since we are interested in direction ratios rather than absolute values, we can say that we intend to walk in the direction which has components $(2, 1)$.

Reference to Fig. 18.1 will show us the starting point for our walk and the direction. We will also see that this direction causes us to head for the constraint where $2\pi r^2 + 2\pi rl = 10$.

Let us reconsider our problem, taking account of this constraint. We rewrite our objective function as:

$$F = V - \frac{0\cdot01}{10 - 2\pi r^2 - 2\pi rl}$$

i.e. $$F = \pi r^2 l - \frac{0\cdot01}{10 - 2\pi r^2 - 2\pi rl}$$

(really we should set up the function

$$F = V - \frac{0\cdot01}{\frac{3}{2}r - l} - \frac{0\cdot01}{l} - \frac{0\cdot01}{10 - 2\pi r^2 - 2\pi rl}$$

in order to take account of all constraints but this would increase the complexity of our arithmetic beyond that which is necessary to understand the method).

Now the desirable direction has components $\partial F/\partial r$ and $\partial F/\partial l$, and

$$\frac{\partial F}{\partial r} = 2\pi rl - \frac{0\cdot01\,(4\pi r + 2\pi l)}{(10 - 2\pi r^2 - 2\pi rl)^2}$$

and

$$\frac{\partial F}{\partial l} = \pi r^2 - \frac{0\cdot01\,(2\pi r)}{(10 - 2\pi r^2 - 2\pi rl)^2}$$

If we evaluate these components at the point $(0\cdot2, 0\cdot2)$ we will see that

$$\frac{\partial F}{\partial r} = 0\cdot08(\pi) \text{ and } \frac{\partial F}{\partial l} = 0\cdot04(\pi)$$

are acceptable approximations so that, as in the unconstrained problem, we will walk to a new point with co-ordinates

$$r = 0\cdot2 + 2t, \quad l = 0\cdot2 + t$$

Substituting this point in our expression for F we find

$$F(t) = \pi(0\cdot2 + 2t)^2\,(0\cdot2 + t) - \frac{0\cdot01}{10 - 2\pi(0\cdot2 + 2t)^2 - 2\pi(0\cdot2 + 2t)(0\cdot2 + t)}$$

$$= \pi(0\cdot008 + 0\cdot2t + 1\cdot6t^2 + 4t^3) - \frac{0\cdot01}{9\cdot5 - 8\cdot8t - 38t^2}$$

To find the best value of t, we optimize $F(t)$.   (i)

$$\frac{\mathrm{d}F(t)}{\mathrm{d}t} = \pi(0\cdot2 + 3\cdot2t + 12t^2) - \frac{0\cdot01\,(8\cdot8 + 76t)}{(9\cdot5 - 8\cdot8t - 38t^2)^2}$$

$$= 0 \text{ for an extreme } F(t)$$

We now see one of the difficulties of hill climbing for we cannot solve t by any analytical method. Trial and error will show that t is less than 0·4 and more than 0·3. We can legitimately take $t = 0·3$ and walk to the point where

$$r = 0·8$$
$$l = 0·5,$$

knowing that we will have improved our position without breaking the constraint.

From this new point, we may repeat the process, calculating the best direction in which to walk and walking as far as possible in that direction without breaking the constraint. We will see eventually that it is desirable to reduce ρ to a number much below 0·01 for a new step because our direction will otherwise become too dominated by the constraint. It will also be realized that if we have to use approximate methods to find t from equation (i), we might as well simplify our working by taking short arbitrary strides and testing for altitude. Most computer solutions would, in fact, choose t without reference to equation (i).

Methods of optimization are not yet used frequently in the design of complex systems because of the difficulty of representing any real system by manageable, algebraic expressions. Some work is being done to design structures of the lightest weight but only in a few cases have practicable results been obtained for real situations.

Hill climbing methods are being used in shipbuilding [18(c)] to define the geometry of structures for which the topology is already determined.

A further application of optimization occurs in circuit design [to 18(d)]. Consider a situation in which a circuit containing L, R and C is required to produce an output voltage V_0, varying in accordance with a previously stated law with input voltage V_i. Given the circuit, V_0 can be calculated as a function of the L, R and C of each component of the network, for a number of expected values of V_i and the sum of the squares of the deviations from the desired values may be used as an objective which is to be minimized, so that L, R and C are eventually chosen, for each component of the system, in such a way that the desired law is obtained as nearly as possible.

One situation in which optimization has been used [18 (e)] is relevant to the pressure switch of Fig. 2.1. Consider the metal bellows of that switch. It can be shown with some approximation that the stress induced in a metal bellows is given by

$$f = At + Bt^2 \qquad (i)$$

where t is the thickness of the metal, A is a function of diameter, deflection and convolution geometry and B is a function of pressure.

Also, the Euler stability is assured if the stiffness K of the bellows meets the requirement

$$K \geqslant \frac{PL}{3} \qquad (ii)$$

where P is the pressure, L the length of the bellows and K is a function of diameter, thickness and convolution geometry.

The volume of material in the bellows may also be taken as

$$V = 6 \cdot 3 \, D_m \, Nnt\rho \, (d + 0 \cdot 3p) \qquad \qquad (iii)$$

where D_m is the mean diameter, N the number of convolutions, n the number of ply, t the thickness of material, ρ the density of material, d the depth of convolution and p the pitch.

If we specify the life of the bellows in terms of load reversals, Fig. 16.11 will tell us the maximum permitted stress level, f^*.

Minimizing the weight of the bellows would therefore require that we

$$\text{minimize} \quad V = D_m \, Nnt\rho \, (d + 0 \cdot 3p)$$

$$\text{subject to} \quad f \leqslant f^*$$

$$\text{and} \quad K \geqslant \frac{PL}{3}$$

This is apparently a Kuhn–Tucker or hill climbing problem. In fact the problem is complicated by the need to use integral numbers of convolutions and thicknesses which are S.W.G. The problem has, however, been solved to an extent which can be used commercially.

Bibliography

18(a) I. Sokolnikoff and R. Redheffer, *Mathematics of Physics and Modern Engineering*, McGraw-Hill, New York, 1958.

18(b) R. Gue and M. Thomas, *Mathematical Methods in Operations Research*, Collier–MacMillan, New York, 1968.

18(c) J. Moe, *Design of Ship Structures by Means of Non-Linear Programming Techniques*, Meddelelse SKB II/M16, Norges Tekniske Hogskole, Trondheim, N.T.H., 1968.

18(d) F. Kuo and J. Kaiser, *System Analysis by Digital Computer*, Wiley, New York, 1966.

18(e) Industrial Systems Engineering Division, University College Swansea, Group Project Report, 1969–1970.

CHAPTER 19
Linear Programming

It is desirable for the student to study at least one optimization technique to a point where he can do simple calculations. For this reason linear programming has been chosen for a fuller treatment than other methods. It may also be commented that in many undergraduate courses there is already some study of methods which depend on the calculus. If we add an introduction to linear programming to the student's repertoire, he is then equipped to consider optimization in greater detail, using methods which draw both from classical techniques and also from the more recently devised programming methods.

The chapter is neither exhaustive nor rigorous in treatment. For more rigour, the student is referred to such texts as Smythe and Johnson, 1966, which gives a satisfying elementary treatment, or Gass [19(a)], which leads to the discussion of the slightly more advanced topics of duality, ranging etc.

There is little writing which directly relates linear models to design but Charnes and Cooper, 1961, is a minor classic which discusses linear economic models.

The early paragraphs in this chapter summarize and simplify the early paragraphs of chapter 18. This is done so that, if desired, Linear Programming may be discussed without reference to other chapters.

LINEAR PROGRAMMING

The basic problem for the designer is to meet the needs of the customer, within his resources and at maximum profit. He has then to maximize a profit function subject to certain constraints. The constraints are provided partly by resources and partly by minimum performance requirements. The system being designed will be defined by a number of parameters (x_i, $i = 1, 2, ..., n$) and the designer's job is to choose values of x_i which will maximize the profit z.

x_i could, for example, be sufficient geometric co-ordinates to define the geometry of the whole of the system, or the lengths of sufficient number of struts and ties to define a structure, or the values of reactances in all the limbs of an electrical network or any number of possible combinations of parameters.

The co-ordinates (x_i) define the system completely and once the system is defined, the profit z to be obtained from it is (in theory) determined, so that z is a function of (x_i).

The resources used and the performance achieved will also be determined when the values (x_i) are fixed. That is, the quantity of any resource used, B_j, must also be a function of (x_i). A performance, P_k, achieved by the system will also (in theory) be determined when the values (x_i) are fixed so that every P_k will be a function of (x_i).

The designer's problem can therefore be stated in the form: choose (x_i) so as to maximize

$$z = z(x_i) \tag{19.1}$$

subject to relationships of the form

$$B_j(x_i) \leqslant b_j \tag{19.2}$$

$$P_k(x_i) \geqslant p_k \tag{19.3}$$

where b_j is the maximum available amount of the jth resource and p_k is the minimum acceptable value of the kth performance requirement.

In a real-life problem, we may find it useful to simplify the problem by restating it. The aircraft designer may try to minimize weight because weight is easier to compute than profit and is assumed to be the most significant term in the profit function. The motor car designer may also minimize weight since he feels that minimizing the weight of a car tends to minimize the cost of its production. The engine designer may attempt to minimize fuel consumption or to maximize the thermodynamic efficiency of a turbine. All these are variants on the same theme; that is, an attempt to maximize profit.

Generally, of course, the functions to be optimized and the expressions for constraints will be very complicated. In the simplest form in which the problem can occur, however, the objective function and constraints are linear.

The linear problem would be

$$\text{Maximize} \quad z = e_1 x_1 + e_2 x_2 + \ldots + e_i x_i + \ldots e_n x_n \tag{19.4}$$

$$\text{Subject to}$$

$$\left. \begin{array}{l} a_{11} x_1 + a_{12} x_2 + \quad \ldots \quad + a_{1i} x_i + \quad \ldots \quad + a_{1n} x_n \leqslant b_1 \\ \cdots\cdots\cdots\cdots\cdots\cdots\cdots\cdots\cdots\cdots\cdots\cdots\cdots \\ a_{j1} x_1 + \quad\quad\quad \ldots \quad + a_{ji} x_i + \quad \ldots \quad + a_{jn} x_n \leqslant b_i \\ \cdots\cdots\cdots\cdots\cdots\cdots\cdots\cdots\cdots\cdots\cdots\cdots\cdots \\ a_{m1} x_1 + \quad\quad\quad \ldots \quad + a_{mi} x_i + \quad \ldots \quad + a_{mn} x_n \leqslant b_m \end{array} \right\} \tag{19.5}$$

$$\left. \begin{array}{l} a_{m+1,\,1} x_1 + \quad\quad\quad\quad\quad \ldots \quad\quad\quad\quad\quad\quad \geqslant p_{m+1} \\ \cdots\cdots\cdots\cdots\cdots\cdots\cdots\cdots\cdots\cdots\cdots\cdots\cdots \\ \cdots\cdots\cdots\cdots\cdots\cdots\cdots\cdots\cdots\cdots\cdots\cdots\cdots \\ a_{m+s,\,1} x_1 + \quad \ldots \quad + a_{m+s,\,i} x_i + \ldots + a_{m+s,\,n} x_n \geqslant p_{m+s} \end{array} \right\} \tag{19.6}$$

Such a problem is capable of solution by analytical methods and, in particular, the simplex algorithm is available to solve it and can readily be used to obtain a computer solution. The solution to the linear problem is called a linear programme.

There are, in fact, few situations where the designer can use linear programming as a basis for solving his major problem. Even simple design problems produce objective functions and constraints that are very far from linear so that most problems are, in fact, solved heuristically. Some work on structures and in other restricted fields of design will eventually yield useful methods based on linear programming, but generally the method is most useful for solving sub-problems in the work of design.

Suitable problems for linear programming are to determine the best product mix or the best use of resources.

Example 1

Let us assume that we are making flags which use red, white and blue cloth, that we have 15 yards of red cloth, 17 yards of white cloth and 16 yards of blue cloth. We can make two types of flag.

Flag A requires 3 yards of red and 3 yards of white cloth.

Flag B requires 4 yards of blue cloth and 2 yards of white cloth.

Each flag A makes a profit of £5 and each flag B makes a profit of £3.

If we make x flags of type A and y flags of type B then our profit $z = 5x + 3y$ but we use $3x$ yards of red cloth,

$$3x + 2y \text{ yards of white cloth}$$
$$\text{and } 4y \text{ yards of blue cloth}$$

as we are limited in our resources to 15, 17 and 16 yards of red, white and blue cloth respectively,

$$3x \quad\quad \leqslant 15$$
$$3x + 2y \leqslant 17$$
$$4y \leqslant 16$$

Our problem can therefore be written:
maximize $z = 5x + 3y$ (i)
subject to

$$3x \quad\quad \leqslant 15 \tag{ii}$$

$$3x + 2y \leqslant 17 \tag{iii}$$

$$4y \leqslant 16 \tag{iv}$$

and $x \geqslant 0$, $y \geqslant 0$ (v)

(these last, non-negative constraints, apply because we obviously cannot make negative numbers of any sort of flag).

If we express constraints (ii) and (iv) graphically we have the lines DC and AB of Fig. 19.1.

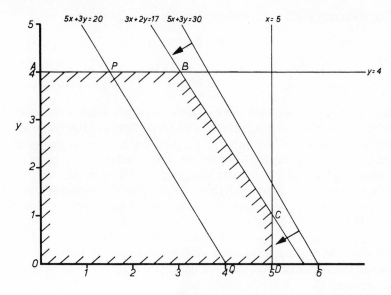

Fig. 19.1

The values of x and y must be such that the point (x, y) lies on the shaded side of both lines, $y = 4$ and $x = 5$.

Constraint (iii) involves only slightly more difficulty for we can see that if we plot the line $3x + 2y = 17$, BC on our diagram, (x, y) must be on the shaded side of this line also and we see that we have drawn a polygon $ABCDO$ within, or on which, our point (x, y) must be if it is to meet all our constraints (remember that constraints (v) mean that the point (x, y) must lie above the x axis and to the right of the y axis). Having drawn our feasible region we have to find a point within it for which z is as large as possible. Now suppose z were 30 then, from (i)

$$30 = 5x + 3y$$

We may plot the line $30 = 5x + 3y$ and we will see that every point on it has the same profit (£30) but that no point on it lies in the feasible region. Suppose $z = 20$, then we could consider and plot the line $20 = 5x + 3y$. Every point on this line gives a profit of £20 and all the points on this line between P and Q lie in the feasible region.

It requires only slight mathematical knowledge to realize that $z = 5x + 3y$ is a family of parallel straight lines, each value of z giving a member of the family. As z is reduced from 30 to 20, the line moves, parallel to itself, in the direction of the arrows. Clearly at some value of z, the line $z = 5x + 3y$ will just touch the feasible region. This will, in fact, happen at the point C so that the point C is that point in the feasible region for which z has the highest value. To find any other point on our moving line that is in the feasible region, would involve reducing z, i.e. reducing profit. To raise z would involve moving our line away from the feasible region so that our profit cannot, in fact, be increased.

Our maximum profit is therefore achieved at the point C, when

$$x = 5$$
$$y = 1$$
$$\text{and } z = 5x + 3y$$
$$= 28$$

It is important to see that our optimum point will be at a corner of the feasible region except in the special case where the family of lines representing our objective function is parallel to a constraint it meets at the optimum point. Even in this situation, however, our answer will lie at any point on the constraint and a corner will be an acceptable optimum point.

When a problem has two variables, we can plot the constraints and the objective function on rectangular axes and solve it in the manner described above. If the problem has three variables it would be feasible, although tedious, to construct a three-dimensional model on three rectangular axes and solve it. In real life, however, a problem is unlikely to have only two or three variables. Two or three hundred unknowns are more likely in an industrial situation.

However, if we have n variables, the geometrical ideas that we used in the two-dimensional case can be carried through to n dimensional geometry.

Consider the general problem in three dimensions, x_1, x_2, x_3. There will be an objective function z, to maximize, say, where

$$z = e_1 x_1 + e_2 x_2 + e_3 x_3 \tag{19.7}$$

subject to a number of constraints such as

$$a_{11} x_1 + a_{12} x_2 + a_{13} x_3 \leqslant b_1 \tag{19.8}$$

$$a_{21} x_1 + a_{22} x_2 + a_{23} x_3 \leqslant b_2 \tag{19.9}$$

$$a_{31} x_1 + a_{32} x_2 + a_{33} x_3 \leqslant b_3 \tag{19.10}$$

$$a_{41} x_1 + a_{42} x_2 + a_{43} x_3 \leqslant b_4 \tag{19.11}$$

and

$$x_1 \geqslant 0, x_2 \geqslant 0, x_3 \geqslant 0 \tag{19.12}$$

We introduce new variables x_4, x_5, x_6, x_7 (all non-negative) so that our constraints may be written

$$a_{11} x_1 + a_{12} x_2 + a_{13} x_3 + x_4 \qquad\qquad = b_1 \tag{19.13}$$

$$a_{21} x_1 + a_{22} x_2 + a_{23} x_3 \qquad + x_5 \qquad = b_2 \tag{19.14}$$

$$a_{31} x_1 + a_{32} x_2 + a_{33} x_3 \qquad\qquad + x_6 \quad = b_3 \tag{19.15}$$

$$a_{41} x_1 + a_{42} x_2 + a_{43} x_3 \qquad\qquad\qquad + x_7 = b_4 \tag{19.16}$$

Equation (19.7) represents a family of planes; each value of z giving us a plane of the family. Equations (19.13), (19.14), (19.15), (19.16) also represent families of planes. Each value of x_4 gives a plane of the family of equation (19.13), values of x_5, x_6, x_7 fix planes in the other families.

If x_4 is positive or zero

$$a_{11}x_1 + a_{12}x_2 + a_{13}x_3 \leqslant b_1$$

and constraint (19.8) is satisfied. This may be interpreted geometrically by saying that if $x_4 > 0$ then the point (x_1, x_2, x_3) lies on the feasible side of the plane

$$a_{11}x_1 + a_{12}x_2 + a_{13}x_3 = b_1$$

if $x_4 = 0$, then the point (x_1, x_2, x_3) lies on the plane (and is therefore feasible) whereas, if $x_4 < 0$, the point (x_1, x_2, x_3) lies on the non-feasible side of the plane. Similar comments apply for the planes of (19.14), (19.15) and (19.16) when x_5, x_6, x_7 are greater than, equal to or less than zero. x_4, x_5, x_6 and x_7 are known as *slack* variables.

The planes

$$x_4 = 0 \text{ or } a_{11}x_1 + a_{12}x_2 + a_{13}x_3 = b_1 \tag{19.17}$$

$$x_5 = 0 \text{ or } a_{21}x_1 + a_{22}x_2 + a_{23}x_3 = b_2 \tag{19.18}$$

$$x_6 = 0 \text{ or } x_{31}x_1 + a_{32}x_2 + a_{33}x_3 = b_3 \tag{19.19}$$

$$x_7 = 0 \text{ or } x_{41}x_1 + a_{42}x_2 + a_{43}x_3 = b_4 \tag{19.20}$$

$$\left. \begin{array}{l} x_1 = 0 \\ x_2 = 0 \\ x_3 = 0 \end{array} \right\} \tag{19.21}$$

therefore between them form a polyhedron which encloses the feasible region. Each of the corners of the polyhedron will be defined by three of the planes (three planes meet at a corner of the polyhedron) so that at each corner of the feasible region, three of the variables $(x_i, i = 1, 2, ..., 7)$ will be zero.

Extending our two-dimensional argument, we will expect our optimum to occur at a corner of the feasible region and so we need consider only those solutions of expressions (19.12), (19.13), (19.14), (19.15) and (19.16) for which three of the variables are zero. There are four equations (19.13), (19.14), (19.15) and (19.16) in seven variables so that we would expect an infinity of solutions. The number of solutions is restricted by (19.12) to the positive octant but still remains an infinity. Provided, however, that our feasible region is a closed polyhedron, when we add the further constraint that three of the variables must be zero, we have a finite, manageable number of solutions.

Now consider solving (19.13), (19.14), (19.15) and (19.16). We may adopt a shorthand method of writing these equations, thus:

$$
\begin{bmatrix}
a_{11} & a_{12} & a_{13} & 1 & 0 & 0 & 0 & b_1 \\
a_{21} & a_{22} & a_{23} & 0 & 1 & 0 & 0 & b_2 \\
a_{31} & a_{32} & a_{33} & 0 & 0 & 1 & 0 & b_3 \\
a_{41} & a_{42} & a_{43} & 0 & 0 & 0 & 1 & b_4
\end{bmatrix}
\tag{19.22}
$$

We are only interested in solutions where three of the variables are zero so that we could consider a solution in which $x_1 = x_2 = x_3 = 0$, then

$$
\begin{aligned}
x_4 &= b_1 \\
x_5 &= b_2 \\
x_6 &= b_3 \\
x_7 &= b_4
\end{aligned}
$$

and we could calculate

$$
z = e_1(0) + e_2(0) + e_3(0)
$$

If we believe that this solution is not the best then we can move to another corner of the feasible region. Observing that, in matrix (19.22) only the 4th, 5th, 6th and 7th columns provide non-zero variables and that these columns are all the possible unit vectors (all entries zero except one, which is unity), it is reasonable to ask if we can change the matrix so that other columns become unit vectors. The set of all the unit vectors is called the basis and the members of the basis are called basic vectors. Let us rearrange the matrix so that the first column is a unit vector and has a unit element in the first row. Since (19.22) is merely a number of equations, it is permissible to divide the first row, throughout, by $a_{11}{}^*$, to yield:

$$
\begin{bmatrix}
1 & \dfrac{a_{12}}{a_{11}} & \dfrac{a_{13}}{a_{11}} & \dfrac{1}{a_{11}} & 0 & 0 & 0 & \dfrac{b_1}{a_{11}} \\[2ex]
a_{21} & a_{22} & a_{23} & 0 & 1 & 0 & 0 & b_2 \\[1ex]
a_{31} & a_{32} & a_{33} & 0 & 0 & 1 & 0 & b_3 \\[1ex]
a_{41} & a_{42} & a_{43} & 0 & 0 & 0 & 1 & b_4
\end{bmatrix}
\tag{19.23}
$$

* Provided that $a_{11} \neq 0$.

Again, since (19.23) represents four equations, we may multiply the first row by a_{21} and subtract the result from the second row, thus:

$$
\begin{bmatrix}
1 & \dfrac{a_{12}}{a_{11}} & \dfrac{a_{13}}{a_{11}} & \dfrac{1}{a_{11}} \ 0 \ 0 \ 0 & \dfrac{b_1}{a_{11}} \\[3ex]
0 & \left\{a_{22}-\dfrac{a_{21}\cdot a_{12}}{a_{11}}\right\} & \left\{a_{23}-\dfrac{a_{21}\cdot a_{13}}{a_{11}}\right\} & \dfrac{-a_{21}}{a_{11}} \ 1 \ 0 \ 0 & b_2-\dfrac{b_1\,a_{21}}{a_{11}} \\[3ex]
a_{31} & a_{32} & a_{33} & 0 \quad 0 \ 1 \ 0 & b_3 \\[2ex]
a_{41} & a_{42} & a_{43} & 0 \quad 0 \ 0 \ 1 & b_4
\end{bmatrix}
\qquad (19.24)
$$

By similar treatment of rows 3 and 4 we obtain

$$
\begin{bmatrix}
1 & \dfrac{a_{12}}{a_{11}} & \dfrac{a_{13}}{a_{11}} & \dfrac{1}{a_{11}} \ 0 \ 0 \ 0 & \dfrac{b_1}{a_{11}} \\[3ex]
0 & \left\{a_{22}-\dfrac{a_{21}\,a_{12}}{a_{11}}\right\} & \left\{a_{23}-\dfrac{a_{21}\,a_{13}}{a_{11}}\right\} & \dfrac{-a_{21}}{a_{11}} \ 1 \ 0 \ 0 & \dfrac{b_2-b_1\,a_{21}}{a_{11}} \\[3ex]
0 & \left\{a_{32}-\dfrac{a_{31}\,a_{12}}{a_{11}}\right\} & \left\{a_{33}-\dfrac{a_{31}\,a_{13}}{a_{11}}\right\} & \dfrac{-a_{31}}{a_{11}} \ 0 \ 1 \ 0 & \dfrac{b_3-b_1\,a_{31}}{a_{11}} \\[3ex]
0 & \left\{a_{42}-\dfrac{a_{41}\,a_{12}}{a_{11}}\right\} & \left\{a_{43}-\dfrac{a_{41}\,a_{13}}{a_{11}}\right\} & \dfrac{-a_{41}}{a_{11}} \ 0 \ 0 \ 1 & \dfrac{b_4-b_1\,a_{41}}{a_{11}}
\end{bmatrix}
\qquad (19.25)
$$

and we see that effectively we have taken out the unit vector from the 4th column and put one in the first column instead. Or, whereas the basic vectors were the 4th, 5th, 6th and 7th columns they are now the 1st, 5th, 6th and 7th.

We can see that a solution of the equations represented by (19.25) is

$$x_1 = \frac{b_1}{a_{11}}$$

$$x_2 = x_3 = x_4 = 0$$

$$x_5 = b_2 - \frac{b_1\,a_{21}}{a_{11}}$$

$$x_6 = b_3 - \frac{b_1\,a_{31}}{a_{11}}$$

$$x_7 = b_4 - \frac{b_1 \, a_{41}}{a_{11}}$$

$$\text{and } z = e_1 \frac{b_1}{a_{11}} + e_2(0) + e_3(0)$$

It is possible to extend the argument to a situation in which we have n variables and m inequalities. Our problem would then be to maximize

$$z = \sum_{i=1}^{n} e_i \, x_i$$

subject to m inequalities of the form

$$\sum_{i=1}^{n} a_{ji} \, x_i \leqslant b_j, \qquad j = 1, 2, ..., m$$

Repeating the above algebra in the more general case we have as our first matrix

$$\begin{bmatrix} a_{11} \, x_1 \; ... \; a_{1n} \, x_n & 1 & 0 & & 0 & b_1 \\ \cdot \; \cdot \; \cdot \; \cdot \; \cdot \; \cdot \; \cdot \; \cdot \; \cdot \; \cdot \; \cdot \; \cdot \; \cdot \; \cdot & & & & \cdot & \\ \cdot \; \cdot \; \cdot \; \cdot \; \cdot \; \cdot \; \cdot \; \cdot \; \cdot \; \cdot \; \cdot \; \cdot \; \cdot \; \cdot & & & & \cdot & b_r \\ a_{r1} \, x_1 & & ... & 1 \; ... \; 0 & & \\ \cdot \; \cdot \; \cdot \; \cdot \; \cdot \; \cdot \; \cdot \; \cdot \; \cdot \; \cdot \; \cdot \; \cdot \; \cdot \; \cdot & & & & \cdot & \\ \cdot \; \cdot \; \cdot \; \cdot \; \cdot \; \cdot \; \cdot \; \cdot \; \cdot \; \cdot \; \cdot \; \cdot \; \cdot \; \cdot & & & \cdot & \cdot & \\ a_{m1} \, x_1 & & ... & & 1 & b_m \end{bmatrix} \qquad (19.26)$$

Without loss of generality we can take matrix (19.26) to be any matrix on the way to finding the optimum objective function and the objective function as

$$z = \sum_{i=1}^{m+n} e_i \, x_i \quad (e_i = 0 \text{ if } x_i \text{ is a slack variable}) \qquad (19.27)$$

Choose r and k such that the $(n+r)$th column is a basic vector and the kth column is not. Observe that the $(n+r)$th column will be a unit vector with unity in its rth row.

One corner of the feasible region is seen from (19.26) to be

$$\left. \begin{array}{l} x_1 = x_2 = ... = x_n \quad = 0 \\ \qquad\qquad x_{n+1} = b_1 \\ \qquad\qquad x_{n+2} = b_2 \\ \qquad\qquad \cdots\cdots \\ \qquad\qquad x_{n+r} = b_r \\ \qquad\qquad \cdots\cdots \\ \qquad\qquad x_{n+m} = b_m \end{array} \right\} \qquad (19.28)$$

and at this corner, from (19.27)

$$z = \sum_{i=1}^{m} e_{n+i} b_i \qquad (19.29)$$

If we now use the same algebra as before to insert the kth column into the basis in place of the $(n+r)$th column we would reach a corner of the feasible region at which

$$\left.\begin{array}{l} x_i' = 0, \; i = 1, 2, \ldots, n, \; i \neq k \\[1ex] \qquad\qquad\qquad i = n+r \\[2ex] x_{n+i}' = b_i - \dfrac{b_r \, a_{ik}}{a_{rk}}, \; i = 1, \ldots, m, \; i \neq r \\[3ex] x_k' = \dfrac{b_r}{a_{rk}}, \\[2ex] \text{where } a_{rk} \neq 0 \end{array}\right\} \qquad (19.30)$$

At this corner, the objective

$$z = \sum_{i=1}^{m+n} e_i x_i'$$

$$= \sum_{\substack{i=1 \\ i \neq r}}^{m} \left(e_{n+i} \left(b_i - \frac{b_r a_{ik}}{a_{rk}} \right) \right) + e_k \frac{b_r}{a_{rk}}$$

$$= \sum_{\substack{i=1 \\ i \neq r}}^{m} e_{n+i} b_i - \sum_{\substack{i=1 \\ i \neq r}}^{m} e_{n+i} \frac{b_r a_{ik}}{a_{rk}} + e_k \frac{b_r}{a_{rk}}$$

$$= z - e_{n+r} b_r - \sum_{\substack{i=1 \\ i \neq r}}^{m} e_{n+i} \frac{b_r a_{ik}}{a_{rk}} + e_k \frac{b_r}{a_{rk}}$$

$$\therefore z - z = \sum_{\substack{i=1 \\ i \neq r}}^{m} e_{n+i} \frac{(-b_r a_{ik})}{a_{rk}} - e_{n+r} b_r + e_k \frac{b_r}{a_{rk}}$$

$$= \frac{b_r}{a_{rk}} \left(e_k - \sum_{i=1}^{m} e_{n+i} a_{ik} \right) \qquad (19.31)$$

Replacing in the basis, the $(n+r)$th vector by the kth vector will therefore lead us to a better solution if $b_r > 0$, $a_{rk} > 0$ and

$$e_k > \sum_{i=1}^{m} e_{n+i} a_{ik} \qquad (19.32)$$

It is also likely to be desirable for

$$e_k - \sum e_{n+i} a_{ik} \text{ to be large} \qquad (19.32)$$

(If we wish to minimize the objective function, it will clearly be necessary for $e_k - \sum e_i a_{ik}$ to be negative and desirable that it should have a large modulus.)

If in the matrix (19.26), all the terms b_j are non-negative, then we will have met the requirements that none of the values x_j are negative.

When we replace the rth vector in the basis, by the kth vector, we must ensure that we do not break the non-negative constraints, so that all our expressions in (19.30) must be non-negative and in particular, those of the form

$$b_i - \frac{b_r a_{ik}}{a_{rk}} \qquad (19.33)$$

and

$$\frac{b_r}{a_{rk}} \qquad (19.34)$$

must be non-negative. If $b_r = 0$, these conditions will be met but when b_r is positive we must ensure that a_{rk} is not negative in order that (19.34) is non-negative. We must then ensure that

$$b_i - \frac{b_r a_{ik}}{a_{rk}} \geqslant 0$$

or

$$\text{or} \quad \frac{b_i}{a_{ik}} \geqslant \frac{b_r}{a_{rk}}$$

where a_{ik} is positive.

It is possible to choose k before specifying r and there is no difficulty in deciding whether a chosen k would increase our objective function. Once having chosen k, however, we have to choose r and when doing so we must ensure that b_r/a_{rk} is the least of the positive ratios b_i/a_{ik}.

The above comments give us the essentials of the 'Simplex' method of linear programming.

Example 2

Consider the simple example which has already been solved graphically:

$$\text{Maximize } z = 5x + 3y \qquad\qquad \text{(i)}$$

subject to

$$3x \quad\;\; \leqslant 15 \qquad\qquad \text{(ii)}$$
$$3x + 2y \leqslant 17 \qquad\qquad \text{(iii)}$$
$$4y \leqslant 16 \qquad\qquad \text{(iv)}$$
$$x, \quad y \leqslant 0 \qquad\qquad \text{(v)}$$

We first introduce slack variables so that (ii), (iii) and (iv) become equations thus:

$$3x \quad\;\; + u \quad\;\;\;\; = 15$$
$$3x + 2y \quad + v \quad\;\; = 17$$
$$4y \quad\;\;\; + w = 16$$

u, v and w are called slack variables and indicate the quantity of each resource that is not used.

If we write this in matrix form we have

$$\begin{pmatrix} 3 & 0 & 1 & 0 & 0 & | & 15 \\ 3 & 2 & 0 & 1 & 0 & | & 17 \\ 0 & 4 & 0 & 0 & 1 & | & 16 \end{pmatrix}$$

It is convenient to construct a tableau about this matrix thus:

Tableau 2.1

		$j = 1$	2	3	4	5		
	e_j	5	3	0	0	0		
Soln	e_i	x	y	u	v	w	b	b_i/a_{ik}
$i = 1$	u	0	③	0	1	0	0	15	5	r
$i = 2$	v	0	3	2	0	1	0	17	$5\frac{2}{3}$
$i = 3$	w	0	0	4	0	0	1	16	∞
$\sum_{i=1}^{3} e_i a_{ij}$		0	0	0	0	0	0		
$e_j - \Sigma e_i a_{ij}$		5	3	0	0	0		

k

Here we have written the appropriate coefficient in the objective function above each variable so that the top row can be interpreted as

$$z = 5x + 3y + 0u + 0v + 0w$$

In the first column, headed soln., we have listed the non-zero variables (i.e. the variables from the basic vectors) so that we can read, directly

$$\left.\begin{array}{l} u = 15 \\ v = 17 \\ w = 16 \\ x, y = 0 \end{array}\right\}$$ as a corner of the feasible region appropriate to our matrix.

In the second column (under e_i) we have written the objective function coefficient appropriate to each non-zero variable.

In the sixth row we form the summation $\sum\limits_{i=1}^{3} e_i a_{ij}$ and in the seventh row we form $(e_j - \sum e_i a_{ij})$. We choose for k, that value of j which makes $(e_j - \sum e_i a_{ij})$ large and positive and clearly our best choice in the example is where $j = 1$. We will therefore attempt to introduce the first column of the matrix (3rd column in the tableau) into the basis. [This applies the rule of expression (19.32)]

Tableau 2.2

	e_j	5	3	0	0	0		
Soln	e_i	x	y	u	v	w	b	b_i/a_{ik}
x	5	1	0	$\frac{1}{3}$	0	0	5	∞
v	0	0	②	-1	1	0	2	1
w	0	0	4	0	0	1	16	4
$\Sigma e_i a_{ij}$		5	0	$\frac{5}{3}$	0	0	25	
$e_j - \Sigma e_i a_{ij}$		0	3	$-\frac{5}{3}$	0	0		

r

k

This column has to replace an rth basic vector and we choose r so that b_r/a_{rk} is the least of the positive ratios b_i/a_{ik}. Having selected the kth column, we are able to determine the values b_i/a_{ik} in the final column of the tableau. The least of the positive ratios, b_i/a_{ik} in our example is 5, giving us the column of coefficients of u as the column to replace.

We can ring the element in the row labelled r and the column labelled k. This is the pivot element and when the kth column becomes a basic vector, the pivot element must become unity.

To obtain unity at the pivot element of Tableau 2.1 we must divide the whole of the row labelled r by 3 to obtain the first row in the matrix of Tableau 2.2.

To obtain a zero in the second element of our new basic vector we must multiply our new first row by 3 and subtract it from the old second row to create the second row in the matrix of Tableau 2.2.

Since the third element of our new basic vector is already zero, the third row can enter our new tableau, unchanged.

We now repeat the procedure of finding the kth column and rth row of Tableau 2.2 and produce Tableau 2.3.

Tableau 2.3

	e_j	5	3	0	0	0		
Soln	e_i	x	y	u	v	w	b	b_i/a_{ik}
x	5	1	0	$\frac{1}{3}$	0	0	5	
y	3	0	1	$-\frac{1}{2}$	$\frac{1}{2}$	0	1	
w	0	0	0	2	-2	1	12	
$\Sigma e_i a_{ij}$		5	3	$\frac{1}{6}$	$\frac{3}{2}$	0	**28**	
$e_j - \Sigma e_i a_{ij}$		0	0	$-\frac{1}{6}$	$-\frac{3}{2}$	0		

In Tableau 2.3 we see that there is no j for which $(e_j - \Sigma e_i a_{ij})$ is positive so that there is no column which it would be profitable to insert into the basis. We have therefore reached the best corner of our feasible region and we can read the optimum point from our tableau:

$$x = 5$$
$$y = 1$$
$$z = 28 \text{ (note that we find the value of } z, \Sigma e_i b_i, \text{ in the } \Sigma e_i a_{ij}$$
row of the b column).

If the problem were to minimize (for example, minimizing cost) we would require that our objective function be reduced in size at every new tableau. Our kth column would therefore be the column for which $(e_j - \Sigma e_i a_{ij})$ is negative and has the largest modulus.

'Greater than' Constraints

A difficulty arises when one or more of our constraints is of the form

$$a_{j1}x_1 + a_{j2}x_2 + \ldots + a_{jn}x_n \geqslant b_j \qquad (19.35)$$

Example 3

Suppose our problem were rewritten with an additional constraint thus:

$$\text{Maximize } z = 5x + 3y \qquad \text{(i)}$$
$$3x \qquad \leqslant 15 \qquad \text{(ii)}$$
$$3x + 2y \leqslant 17 \qquad \text{(iii)}$$
$$4y \leqslant 16 \qquad \text{(iv)}$$
$$x + 3y \geqslant 9 \qquad \text{(v)}$$
$$x, y \geqslant 0 \qquad \text{(vi)}$$

We would be tempted to rewrite the inequalities as:

$$3x + u \qquad\qquad = 15 \qquad \text{(vii)}$$
$$3x + 2y + v \qquad = 17 \qquad \text{(viii)}$$
$$4y \quad +w \quad = 16 \qquad \text{(ix)}$$
$$x + 3y \qquad -s = 9 \qquad \text{(x)}$$

to give a matrix

$$\begin{pmatrix} 3 & 0 & 1 & 0 & 0 & 0 & | & 15 \\ 3 & 2 & 0 & 1 & 0 & 0 & | & 17 \\ 0 & 4 & 0 & 0 & 1 & 0 & | & 16 \\ 1 & 3 & 0 & 0 & 0 & -1 & | & 9 \end{pmatrix} \qquad \text{(xi)}$$

but this would give an initial solution

$$x = 0, \; y = 0, \; y = 15, \; v = 17, \; w = 16$$

$$\text{and} \quad s = -9 \qquad \text{(xii)}$$

which would be absurd since s must be assumed to be positive if the inequality (v) is not to be broken.

One method of dealing with this situation is to introduce artificial, slack variables into the 'greater than' inequalities so that our constraint (x) will become:

$$x + 3y - s + t = 9 \qquad \text{(xiii)}$$

Now slack variables usually have a physical significance but the artificial slack variable, t, has been inserted only as a mathematical trick to enable us to start solving the problem. We will have to ensure that t does not appear in the final solution and it is possible to do this by rewriting the objective function as:

$$z = 5x + 3y - Mt \qquad \text{(xiv)}$$

where M is a very much larger number than any other which we will encounter in the problem. Since, in the solution, some of our variables will be zero, then the form of (xiv) will ensure that one of those zero variables is t.

Consider the tableaux of this problem:

Tableau 3.1

	e_j	5	3	0	0	0	0	$-M$		
Soln	e_i	x	y	u	v	w	s	t	b	
u	0	3	0	1	0	0	0	0	15	
v	0	3	2	0	1	0	0	0	17	$8\frac{1}{2}$
w	0	0	4	0	0	1	0	0	16	4
t	$-M$	1	③	0	0	0	-1	1	9	3 r
		$-M$	$-3M$	0	0	0	M	$-M$	$-9M$	
		$5+M$	$3+3M$	0	0	0	$-M$	0		

k

We see that the insertion of the artificial slack variable has enabled us to write down a tableau with an initial feasible solution.

We can proceed from this tableau in the usual way thus:

Tableau 3.2

	e_j	5	3	0	0	0	0	$-M$	
Soln	e_i	x	y	u	v	w	s	t	b
u	0	3	0	1	0	0	0	0	15
v	0	$\left(\frac{7}{3}\right)$	0	0	1	0	$\frac{2}{3}$	$-\frac{2}{3}$	11
w	0	$-\frac{4}{3}$	0	0	0	1	$\frac{4}{3}$	$-\frac{4}{3}$	4
y	3	$\frac{1}{3}$	1	0	0	0	$-\frac{1}{3}$	$\frac{1}{3}$	3
		1	3	0	0	0	-1	1	9
		4	0	0	0	0	1	$-M-1$	

Right-hand annotations:
- u row: 5
- v row: $\frac{33}{7}$ r
- w row: $-$
- y row: 9

k

But in this tableau, an acceptable value for t would be zero. The matrix in the centre of this tableau represents a reformulation of our equations (vii), (viii), (ix), (x) without the coefficients of t occurring in the basis. We may, therefore, omit terms containing t from further consideration. We simply proceed with the t column omitted from the tableau thus:

Tableau 3.3

Soln	e_j	5	3	0	0	0	0		
Soln	e_i	x	y	u	v	w	s		b
u	0	3	0	1	0	0	0		15
v	0	$\left(\frac{7}{3}\right)$	0	0	1	0	$\frac{2}{3}$		11
w	0	$-\frac{4}{3}$	0	0	0	1	$\frac{4}{3}$		4
y	3	$\frac{1}{3}$	1	0	0	0	$-\frac{1}{3}$		3
		1	3	0	0	0	-1		9
		4	0	0	0	0	1		

5

$\frac{33}{7}$ r

—

9

k

Tableau 3.4 is the final tableau since the final row has no positive values. Our solution is thus

$$z = \frac{195}{7} = 27\frac{6}{7}$$

$$\text{where } x = \frac{33}{7} = 4\frac{5}{7}$$

$$\text{and} \quad y = \frac{10}{7} = 1\frac{3}{7}$$

This can be confirmed graphically and Fig. 19.2 shows the feasible region and a member of the family of objective lines so that it is clear that the optimal solution to the problem will be found at the indicated corner of the feasible region.

Tableau 3.4

		5	3	0	0	0	0		
Soln	e_i	x	y	u	v	w	s		b
u	0	0	0	1	$-\frac{9}{7}$	0	$-\frac{6}{7}$		$\frac{6}{7}$
x	5	1	0	0	$\frac{3}{7}$	0	$\frac{2}{7}$		$\frac{33}{7}$
w	0	0	0	0	$\frac{4}{7}$	1	$\frac{12}{7}$		$\frac{72}{7}$
y	3	0	1	0	$-\frac{1}{7}$	0	$-\frac{3}{7}$		$\frac{10}{7}$
		5	3	0	$\frac{12}{7}$	0	$\frac{1}{7}$		$\frac{195}{7}$
		0	0	0	$-\frac{12}{7}$	0	$-\frac{1}{7}$		

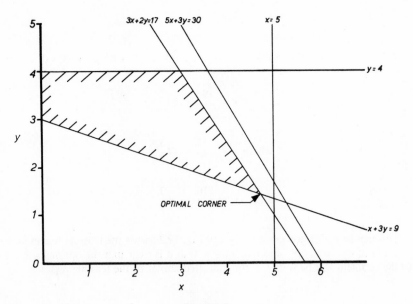

Fig. 19.2

Although the examples that we have chosen to do by the Simplex method have, in fact, been two-dimensional, the method can be used, however many variables we have and however many constraints.

Feasible Region Not Closed

Difficulties do sometimes arise in Linear Programming problems. One difficulty may be that the feasible region is not a closed region.

Example 4

Suppose, for example, we had the problem:

$$\text{Maximize} \quad z = 5x + 3y \tag{i}$$
$$\text{Subject to} \quad x + 3y \geqslant 9 \tag{ii}$$
$$3x + 2y \geqslant 17 \tag{iii}$$

Sketching this problem graphically (Fig. 19.3), we see that the feasible region permits us to raise x and y without limit so that there is no finite solution to the problem.

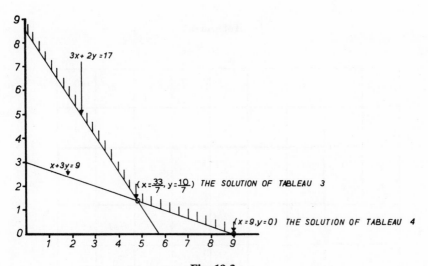

Fig. 19.3

If we attempt to solve the problem by the Simplex method, however, we will have the following sequence of tableaux:

Tableau 4.1

		5	3	0	$-M$	0	$-M$		
		x	y	s	t	s'	t'	b	
t	$-M$	1	③	-1	1	0	0	9	$3\ r$
t'	$-M$	3	2	0	0	-1	1	17	$8\frac{1}{2}$
		$-4M$	$-5M$	M	$-M$	M	$-M$	$-26M$	
		$5+4M$	$3+5M$	$-M$	0	$-M$	0		

k

Tableau 4.2

		5	3	0		0	$-M$		
		x	y	s		s'	t'	b	
y	3	$\frac{1}{3}$	1	$-\frac{1}{3}$		0	0	3	9
t'	$-M$	$\left(\frac{7}{3}\right)$	0	$\frac{2}{3}$		-1	1	11	$\frac{33}{7}\ r$
		$1-\dfrac{7M}{3}$	3	$-1-\dfrac{2M}{3}$		M	$-M$	$9-11M$	
		$4+\dfrac{7M}{3}$	0	$1+\dfrac{2M}{3}$		$-M$	0		

k

Tableau 4.3

		5	3	0		0		
		x	y	s		s'		b
y	3	0	1	$-\frac{3}{7}$		$\left(\frac{1}{7}\right)$		$\frac{10}{7}$
x	5	1	0	$\frac{2}{7}$		$-\frac{3}{7}$		$\frac{33}{7}$
		5	3	$\frac{1}{7}$		$-\frac{12}{7}$		$\frac{195}{7}$
		0	0	$-\frac{1}{7}$		$\frac{12}{7}$		

$$k$$

Tableau 4.4

		5	3	0		0		
		x	y	s		s'		b
s'	0	0	7	-3		1		10
x	5	1	3	-1		0		9
		5	15	-5		0		$\underline{45}$
		0	-12	5		0		

Tableau 4.3 is the first tableau in which the artificial slack variables do not appear and it is possible to plot the solution of this tableau graphically in Fig. 19.3.

Tableau 4.4 indicates that we would like to insert s into the basis but since both the coefficients in the s column are negative, we cannot find a pivot element. We can see the significance of this if we write down the equations represented by the matrix of Tableau 4.4

$$7y - 3s + s' = 10 \qquad \text{(iv)}$$
$$1x + 3y - s = 9 \qquad \text{(v)}$$

We may change the subject of each of these equations thus:

$$s' = 10 - 7y + 3s \qquad \text{(vi)}$$
$$x = 9 - 3y + s \qquad \text{(vii)}$$

We could clearly keep $y = 0$ while increasing s without limit, to increase x without limit and hence increase z without limit.

This arises because all the coefficients of s are negative in Tableau 4.4 so if we have the situation where the objective function is unlimited, we will reach a tableau from which it is desirable to proceed but because of negative coefficients we are unable to find a pivot element.

Contradictory Constraints

A further problem arises when the constraints are contradictory.

Example 5

Consider, for example, the problem:

$$\begin{array}{llr}
\text{Maximize} & z = 5x + 3y & \text{(i)} \\
\text{Subject to} & 3x \quad \geqslant 15 & \text{(ii)} \\
& 4y \quad \geqslant 16 & \text{(iii)} \\
& 3x + 2y \leqslant 17 & \text{(iv)} \\
& x, y \quad \geqslant 0 &
\end{array}$$

If we draw the constraints, as in Fig. 19.4, we see that there is no feasible region since a point (x, y) cannot satisfy all the constraints.

If, however, we attempt to solve the problem by the Simplex method, we create the following equations:

$$\begin{array}{llr}
\text{Maximize} & z = 5x + 3y - Mt - Mt' & \text{(i)} \\
\text{Subject to} & 3x \quad -s + t \quad = 15 & \text{(v)} \\
& 4y \quad -s' + t' = 16 & \text{(vi)} \\
& 3x + 2y + u \quad = 17 & \text{(vii)}
\end{array}$$

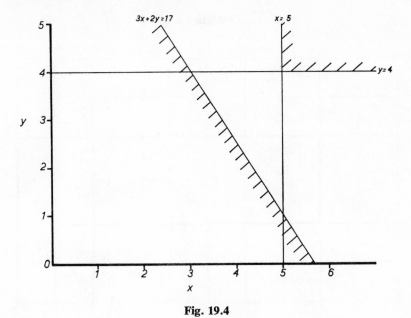

Fig. 19.4

and the sequence of tableaux is:

Tableau 5.1

		5	3	0	0	$-M$	0	$-M$	
		x	y	u	s	t	s'	t'	b
t	$-M$	3	0	0	-1	1	0	0	15
t'	$-M$	0	4	0	0	0	-1	1	16
u	0	3	2	1	0	0	0	0	17
		$-3M$	$-4M$	0	M	$-M$	M	$-M$	$-31M$
		$5+3M$	$3+4M$	0	$-M$	0	$-M$	0	

Tableau 5.2

		5	3	0	0	$-M$	0		
		x	y	u	s	t	s'		b
t	$-M$	3	0	0	-1	1	0		15
y	3	0	1	0	0	0	$-\frac{1}{4}$		4
u	0	3	0	1	0	0	$\frac{1}{2}$		9
		$-3M$	3	0	M	$-M$	$-\frac{3}{4}$		$12-15M$
		$5+3M$	0	0	$-M$	0	$\frac{3}{4}$		

Tableau 5.3

		5	3	0	0	$-M$	0		
		x	y	u	s	t	s'		b
t	$-M$	0	0	-1	-1	1	$-\frac{1}{2}$		6
y	3	0	1	0	0	0	$-\frac{1}{4}$		4
x	5	1	0	$\frac{1}{3}$	0	0	$\frac{1}{6}$		3
		5	3	$M+\frac{5}{3}$	M	$-M$	$\frac{M}{2}+\frac{1}{12}$		
		0	0	$-M-\frac{5}{3}$	$-M$	0	$-\frac{M}{2}-\frac{1}{12}$		

Tableau 5.3 is an apparent final solution since all terms in the final row are zero or negative. Unfortunately one of the artificial slack variables remains in the basis and is hence non-zero. Since we have found a corner of the feasible region only when the artificial slack variables have disappeared from the tableaux, the fact that Tableau 5.3 contains an artificial slack variable means that we have been unable to find a feasible solution—not surprising since none exists.

Conflicting constraints will be shown, in the Simplex method, by our inability to obtain a feasible solution which does not contain artificial variables.

Degeneracy

One of the most interesting difficulties arises through degeneracy. In the degenerate case, one of the corners of the feasible region is overdefined.

Example 6

Consider the problem

$$\text{Maximize} \quad z = 5x + 3y \tag{i}$$
$$\text{Subject to} \quad 3x \quad \leqslant 15 \tag{ii}$$
$$3x + 2y \leqslant 17 \tag{iii}$$
$$4y \leqslant 16 \tag{iv}$$
$$x - y \leqslant 5 \tag{v}$$
$$x, y \geqslant 0 \tag{vi}$$

If we sketch the feasible region, we obtain Fig. 19.5.

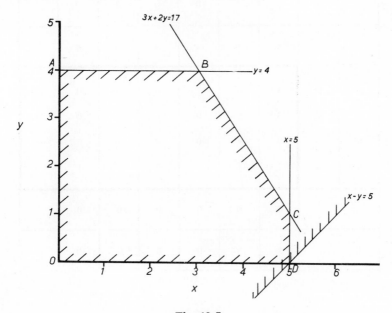

Fig. 19.5

In Fig. 19.5 we see that the feasible region is exactly the same as if we had not included constraint (v), and at the point D, three constraint lines intersect. From Example 1, we know that the solution will be

$$z = 28 \text{ when}$$
$$x = 5$$
$$\text{and} \quad y = 3$$

but if we attempt a Simplex solution of the problem we first rewrite the constraints as equations with slack variables:

$$3x \quad +u \qquad\qquad = 15 \qquad\qquad \text{(vii)}$$
$$3x+2y \quad +v \qquad\quad = 17 \qquad\qquad \text{(viii)}$$
$$4y \qquad +w \quad = 16 \qquad\qquad \text{(ix)}$$
$$x- \ y \qquad\qquad +q = \ 5 \qquad\qquad \text{(x)}$$

Our sequence of tableaux becomes:

Tableau 6.1

		5	3	0	0	0	0	
		x	y	u	v	w	q	b
u	0	3	0	1	0	0	0	15
v	0	3	2	0	1	0	0	17
w	0	0	4	0	0	1	0	16
q	0	1	−1	0	0	0	1	5
		0	0	0	0	0	0	
		5	3	0	0	0	0	

In Tableau 6.1, there is no difficulty in deciding that the x column should be put into the basis but two elements, the first and fourth, have equal claim to be pivot elements. For simplicity, we arbitrarily choose the fourth element as pivot and proceed to Tableau 6.2.

Tableau 6.2

		5	3	0	0	0	0	
		x	y	u	v	w	q	b
u	0	0	3	1	0	0	-3	0
v	0	0	5	0	1	0	-3	2
w	0	0	4	0	0	1	0	16
x	5	1	-1	0	0	0	1	5
		5	-5	0	0	0	5	25
		0	8	0	0	0	-5	

Because of the competing rows in the first tableau, we now have a zero in the b column of Tableau 6.2. The corner of the feasible region that is represented by Tableau 6.2 is

$$x = 5, \ y = 0, \ u = 0, \ v = 2, \ w = 16, \ q = 0.$$

We know that the non-basic variables will be zero but in this case we also have a basic variable, u, which is zero. This is clearly caused by the fact that we are at the point D, of the feasible region and this point lies, not on two constraints but three.

If we continue:

Tableau 6.3

		5	3	0	0	0	0	
		x	y	u	v	w	q	b
y	3	0	1	$\frac{1}{3}$	0	0	-1	0
v	0	0	0	$-\frac{5}{3}$	1	0	2	2
w	0	0	0	$-\frac{4}{3}$	0	1	4	16
x	5	1	0	$\frac{1}{3}$	0	0	0	5
		5	3	$\frac{8}{3}$	0	0	-3	25
		0	0	$-\frac{8}{3}$	0	0	3	

We see that the new solution of Tableau 6.3 is the point where $x = 5$, $y = 0$, $u = 0$, $v = 2$, $w = 16$, $q = 0$.

This is still the point D, of the diagram. y has replaced u in the basis but y is nevertheless, zero.

Continuing:

Tableau 6.4

		5	3	0	0	0	0	
		x	y	u	v	w	q	b
y	3	0	1	$-\frac{1}{2}$	$\frac{1}{2}$	0	0	1
q	0	0	0	$-\frac{5}{6}$	$\frac{1}{2}$	0	1	1
w	0	0	0	2	-2	1	0	12
x	5	1	0	$\frac{1}{3}$	0	0	0	5
		5	3	$\frac{1}{6}$	$\frac{3}{2}$	0	0	28
		0	0	$-\frac{1}{6}$	$-\frac{3}{2}$	0	0	

Tableau 6.4 is the final tableau and gives us the answer we expect.

Degeneracy need not cause us embarrassment and will not often do so if programmes are worked manually. We notice, however, that degeneracy leads us to a situation in which we have to choose between competing pivot elements. It is possible, if care is not taken, to cycle through a number of tableaux without reaching the solution, but in manual solutions this would be obvious and common sense would dictate better choices of pivots. If the solution is being obtained by means of a computer program, common sense cannot be called up midway through the problem and precautions have to be taken to prevent cycling [19(a)].

The problems that we have considered above have all been simple in that they have contained few variables and few constraints. In real life we are likely to have problems with several hundred variables and several hundred constraints and no-one would consider solving them manually. Problems can readily be solved on a computer and most computer firms have programs available for linear programming so that it is not even necessary to write a program (although it is necessary to understand the ideas and nomenclature used).

The commonest uses of linear programming are determining the mix of products that will be made in a factory of known resources (the type of problem already described) or the blending of resources to make the most profitable products (as, for example, in the blending of oil distillates into petrols). The Simplex method may be developed to give more information than merely the optimum value of the objective and the required value of each variable to achieve that objective. It is possible for example, to find the limits over which each coefficient in the objective function may range before the optimal values of the variables change. It is possible, too, to determine the amount by which each resource or performance requirement may be changed without changing the optimal values of the variables. Most commercially-available computer programs will provide such information.

Exercises

1. Solve the following linear programme,
   (i) graphically,
and
   (ii) by the Simplex method.
   Relate each tableau to a point on the graph.

$$\begin{aligned}
\text{Maximize} \quad & 2x + 3y \\
\text{Subject to} \quad & 2x + 7y \leqslant 42 \\
& x \quad\ \leqslant 7 \\
& 4x + 5y \leqslant 40 \\
& x, y \geqslant 0
\end{aligned}$$

2. Solve the following linear programme,
   (i) graphically
and
   (ii) by the Simplex method.
   Relate each tableau to a point on the graph.

$$\begin{aligned}
\text{Minimize} \quad & x + y \\
\text{Subject to} \quad & x \qquad\ \geqslant 1 \\
& \quad\ y \geqslant 1 \\
& 2x + 3y \geqslant 11 \\
& x, y \quad \geqslant 0
\end{aligned}$$

3. Maximize $z = 3x_1 + 4x_2 - x_3 + 2x_4$

$$\begin{aligned}
\text{Subject to} \quad & 3x_1 - x_2 + 2x_4 \leqslant 2 \\
& x_1 + 4x_2 + 2x_3 + x_4 \geqslant 6 \\
& x_1 + x_2 - x_4 \ \leqslant 1 \\
& x_j \geqslant 0 \quad (j = 1, 2, 3, 4).
\end{aligned}$$

4. Solve the following linear programme by the Simplex method.

$$\begin{aligned}
\text{Maximize} \quad & 19x + 10y \\
\text{Subject to} \quad & y \leqslant 3 \\
& x \quad\ \leqslant 4 \\
& x + \ y \leqslant 5 \\
& 2x + \ y \leqslant 8 \\
& x + 4y \geqslant 4 \\
& x, \quad y \geqslant 0
\end{aligned}$$

Comment on the tableaux.
Confirm your comments by solving the programme graphically.

5. Attempt to solve the following linear programme by the Simplex method.

Maximize $x+2y$
Subject to $x- y \leqslant 2$
$3x- y \geqslant 3$
$x, \quad y \geqslant 0$

Comment on the tableaux.
Confirm your comments by drawing the feasible region and a line of constant profit.

6. Attempt to solve the following linear programme by the Simplex method.

Minimize $x+2y$
Subject to $2x+ y \leqslant 2$
$x+2y \leqslant 2$
$x+ y \geqslant 3$
$x, \quad y \geqslant 0$

Comment on the tableaux.
Confirm your comments by drawing the feasible region and a line of constant cost.

7. A firm makes two products, A and B. Each unit of A makes a profit of £6; each unit of B makes a profit of £4. Each unit of A requires 2 tons of ore X and 3 tons of ore Y while each unit of B requires 3 tons of ore X and 2 tons of ore Y.

In order to maintain the goodwill of an important customer it is imperative that the total number of units produced is not less than three in any one week. Only 30 tons of X and 24 tons of Y are available in any week.

Formulate the problem of determining the product mix for maximum profit, as a linear programme. Solve the linear programme without using a graph.

Comment on the final result.

8. A small firm buys castings, forgings and bar which it machines and assembles into two types of finished goods for sale. The operations available within the firm are turning, at £4 per hour, grinding, at £5 per hour and fitting at £3 per hour. The raw materials for one unit of product A cost 50 p and the finished part sells for 125 p while the raw materials for one unit of product B cost 85 p and the finished part sells for 150 p. All three operations are required for each product. A is turned at 20 parts per hour, ground at 27 parts per hour and fitted at 15 parts per hour. B is turned at 25 parts per hour, ground at 22 parts per hour and fitted at 35 parts per hour.

There is a total of 40 hours turning, 40 hours grinding and 40 hours fitting available in any week.

Set up and solve a linear programme to determine how many units of A and how many units of B to make in a week.

9. Girls work in a shop in a factory making plastic brushes.

The three operations considered are filling brushes, assembling and packing. In one day there are available

20 hours of filling
10 hours of assembling
and   25 hours of packing

Each brush, A, requires 2·4 min of filling and 5·0 min of packing.

Each brush, B, requires 3·0 min of filling and 2·5 min of assembly.

Each brush, C, requires 2·0 min of filling, 1·5 min of assembly and 2·5 min of packing.

The profit is 30 p, 35 p, and 25 p on each brush A, B and C respectively.

What is the best distribution of brushes A, B and C to manufacture?

Bibliography

19(a) S. Gass, *Linear Programming, Methods and Applications*, McGraw-Hill, New York, 1958.

Further Reading

W. Smythe and L. Johnson, *Introduction to Linear Programming*, Prentice-Hall, Englewood Cliffs N.J., 1966.

A. Charnes and W. Cooper, *Management Models and Industrial Applications of Linear Programming*, Wiley, New York, 1961.

CHAPTER 20
The Man/Machine Interface

Work design and ergonomics are really disciplines in their own right and worthy of deep and prolonged study but, as with many other topics, the designer must be aware of them and be able to direct the work of specialists. Unfortunately, the work of many designers suggests that they are not even aware of the men in the systems they design. Even without specialist knowledge, many designs could be improved by awareness of the human element and the application of common sense.

This chapter is not intended to do more than demonstrate the problems to intending designers who will, in general, be receiving no further instruction in either work design or ergonomics. The exercises at the end of the chapter reflect this in that they are intended to stimulate discussion rather than serve as numerical examples.

Deeper reading is readily available and typical useful texts are *The Industrial Engineering Handbook* [20(a)] and Murrell [20(d)].

Many designers are so preoccupied with ensuring that in some design case the system gives the required performance that they ignore important contributions to the cost of the system. It is important for the designer to predict the performance of the system but it is also important to predict the cost of achieving that performance.

One of the major contributions to the cost of a system is the cost of labour involved in manufacture. Each detail and assembly will be drawn and from the drawings the planners will decide what operations will be used to manufacture that detail or assembly and in what order the operations will be carried out. It is often the case that the time of each operation will be predicted and that the predicted time of each operation will form the basis of the method by which the operative is paid.

Obviously the shorter, easier and fewer the operations the less will be the labour involved and the cheaper will be the cost of manufacture. When the planner predicts the time that each operation will take, he must predict the movements that the operative must necessarily make and predict the time that each movement will take. If he cannot do this, the cost of an operation may have to be determined by timing it. The designer is not usually involved directly in the devising of payment schemes but it is clear that he must consider the operations involved in manufacture in some detail if he is to ensure that unnecessary expense is not incurred. Since a common fault of designers is to draw components which cannot be made at all, there is much to be gained if they will approach the problems of labour in manufacture with at least commonsense, if not specialist knowledge.

Many methods exist for determining the time taken by an operation. The most obvious and commonly used method is to use a stopwatch to record the actual time taken by a man to perform the operation in question. Unfortunately this method is rarely of use to a designer since it requires that drawings and manufacturing instructions are already available, not normally the case during the design process. Usually work-study consists of finding the best and cheapest way of making an already defined component, but the designer's job is the reverse of this. His work consists of defining the geometry of the part so that, when made by the best available methods, it will be cheaper than any other part of a different geometry which will perform the same function. As in many other situations, the designer has to predict what will happen.

Predictive methods for determining the time to be taken to perform an operation are usually grouped together as Predetermined Motion Time Systems (PMTS). Basically a PMT System breaks down an operation into a number of basic movements (typically REACH, MOVE, GRASP, POSITION, BODY MOTION) [ref. 20(a)] and each movement will be associated with a distance and degree of difficulty. For each method, tables have been prepared to show the time taken by any basic movement, given the distance and degree of difficulty. Typical tables are given (for MTM) in ref. 20(a) while a British PMTS is described, with tables, in ref. 20(b).

A skilled man can predict the time required to perform an operation with acceptable accuracy (a student with some training in MTM predicted the times of operation in the assembly of part of a washing machine to within 5% of the measured times). Normally a designer does not train himself to a high level of skill in any PMTS since he can rely on the availability of specialists, but he must have an appreciation of what is involved.

Generally the predicted time of an operation will be in Standard Minutes. The British Standard Institute defines Standard Performance as 'The rate of output which qualified workers will naturally achieve without overexertion, as an average over the working day or shift provided they know and adhere to the specified method and provided they are motivated to apply themselves to their work.' Standard Performance is denoted by 100 on the B.S. scale and corresponds with 60 Standard Minutes of work in an hour. An operative working at a lower rate to produce, say, half as much output as if he were giving Standard Performance would be rated at 50 on the B.S. scale and would do 30 Standard Minutes of work in an hour. As an incentive it is common to encourage a man to earn time and a third by allowing 80 min to do 60 Standard Minutes' work and scales other than the B.S. 0/100 scale exist.

Frequently used is the 60/80 scale which gives a performance rating of 80 to Standard Performance in the expectation that 80 min will be allowed for 60 min work. That is, a man rated at 80 on the 60/80 scale (100 on the B.S. scale) would be expected to earn 'time and a third' whereas a man rated at 60 would be expected to earn the rate for the job. The various scales used reflect different approaches to the need to rate an operative's work when timing it and generally they will not interest a designer except to the extent that he must understand the terms in which information is supplied to him.

What is certainly needed is that the designer should devise a procedure by which his system can be built and, at least in imagination, predict the operations necessary and ensure that they are feasible. If the operations are either not feasible or are unnecessarily clumsy then part of the design must be reconsidered and possibly the geometry changed. It is perhaps unfortunate that the designer is not usually considered to be directly responsible for the shop-floor manufacturing instructions since he would then be compelled to establish the manufacturing procedure. If it were possible to generalize about the knowledge required by a designer, one would say that in a highly technical industry (say for example, the aerospace industry) the designer would be expected to concern himself with feasibility and broad comparisons of manufacturing methods, whereas in mass-production industries (say for example, the production of motor cars or domestic consumer durables) the designer would be required to have detailed knowledge of work design and be able to forecast the precise cost of manufacturing operations.

On one occasion an order for a comparatively simple, turned part was lost to a competitor who quoted a price lower by 10%. It was subsequently possible for the unsuccessful designer to compare the drawings of both parts and he discovered that while they looked similar, the dimensions were such that his own design required that the part be removed from the chuck, turned round and recentred before machining could be completed, while the competitor had designed a part which could be completely machined once centred in the chuck. The extra operation of removing from the chuck, turning round and recentring accounted for the price difference.

The above example would not in fact have required a detailed PMTS analysis to show the design weakness. The fault would have been shown if a simple list of manufacturing operations had been made by the designer. If we were to consider the production of washing machines, however, we would see that, in technical content, one product does not differ greatly from another. It is likely that a successful washing machine will sell by the hundred thousand so that, although tooling is expensive, its cost per machine will be small and the most significant cost will be the labour in each machine. Under these circumstances, the designer must consider manufacturing methods and operation times from the conception stage of the design.

Frequently, more than one resource is required to perform an operation.

Taking a very simple case of a man turning bar to a required diameter we could have the situation shown in Fig. 20.1. We see, as we have already seen in Chapter 14, that, with more than one resource required, idleness is almost inevitable. This leads us to ask whether alternative employment can be found for the man (and machine, although in the case illustrated, a simple centre lathe, this would be difficult) during his idle periods. Can he profitably look after several machines? Can the parts be so designed that the man's idle time is exactly used by another operation?

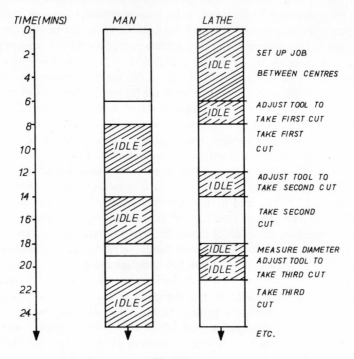

Fig. 20.1

A particular situation in which more subtle consideration of the labour content of manufacturing operations is necessary, occurs when line production is contemplated. Clearly, if we have a sequence of operations to be carried out on a line of components and if all the operations require different times, there will be some difficulty in ensuring that every operative is fully employed. Suppose, for example, we have a washing machine assembly line in which operator A takes 40 seconds to make a weld and his operation is followed by one in which operator B takes 30 seconds to make a further weld. If we have only one operator A and one operator B then clearly B will be idle for a quarter of his time. We could have more operators like A than like B but to achieve complete balance might well embarrass other operations in the same line. Usually it is the planner's duty to balance a line and usually the balance is achieved by a careful breakdown of operations and a careful choice of operative ratios, but an intelligent designer should consider the operations that his drawings demand from the point of view of line design.

Another area in which careful consideration by the designer of the operations involved may save money, is that of test, inspection and adjustment. It may be well worth while to add complication to a design to enable the operative to make adjustments easily, quickly and cheaply.

Consider the pressure switch of Fig. 2.1. Ultimately this has to be adjusted so that the contacts break at a predetermined pressure. This adjustment could be achieved in several ways. One apparently simple way would be to shim under the bellows until the contacts break at the required pressure, but while this would make for a simple and rugged assembly, the setting procedure would be long and tedious. It is quite probable that it would cost much less, in the long run, to build-in a continuously variable, sealed adjustment to the position of the contacts or beam. With such an adjustment available, the operative could apply the appropriate pressure to the switch and simply vary the adjustment until the contacts broke. This contact break could easily be observed if a lamp were put in circuit. Such a method would be quicker than the use of shims with all the stripping and reassembly required. Which method to use would have to be decided at the drawing stage.

Human capabilities must be considered if the designer is to design a system which can be manufactured cheaply but it is rarely the designer's job to provide a good environment for the manufacturing operative. The designer must be considered wholly responsible, however, for providing a good environment for those who will operate his system, and lack of consideration of the operator is still frequent enough to deserve comment. One or two motor cars are reputedly comfortable to drive and the manufacturers use this quality as a selling point. It is very noticeable however, that those cars which possess uncomfortable features are not made cheaper thereby. They are less comfortable than other cars because the designer did not think about the particular feature which causes discomfort.

A survey [20(c)] to discover the range of size and shapes of car interiors was recently carried out on 100 motor cars on which reports had been published in the motoring press. All the cars were available on the British market between 1960 and 1962. The survey made it very clear that 'there is no universal man for whom designers are catering'. For example, 'one manufacturer finds that 15 inches is sufficient distance between the front of the seat and the pedal' while 'another manufacturer provides 25 inches'. 'Examination of the data showed that while price and maximile speed were closely related, no dimension (dimensions were those relating to drive position, position of controls etc.) correlated with price with the exception of seat depth'.

If the designer is making space in his system for a man (or woman) then it is sensible for him to look at a man's dimensions. He must also ask himself whether the space should permit anyone to fit into the system or whether some restriction is desirable. Usually a car-driving position, an aeroplane seat, a work bench, a control console, etc. would be designed to permit easy use by a stated percentage of the population. This is a reasonable procedure because it would be an impossible task to design say, a seat or a console which would accommodate any member of the population, from the smallest to the largest, whereas a seat which will accommodate say 95% of the popu-

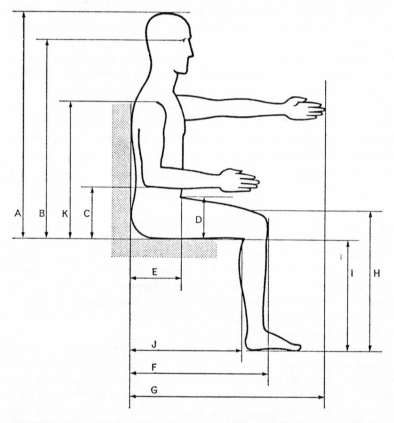

Fig. 20.2 Some of the body dimensions relevant to the design of car interiors about which information is already available. (With acknowledgments to Prof. N. S. Kirk, University of Technology, Loughborough.)

lation will require us to consider only about 1·6 standard deviations from each of the mean dimensions.

The following table, taken from reference 20(c) lists the dimensions of the 5th percentile woman and of the 95th percentile man which are important in the design of a car driver's seat.

Body Size Data (unclothed)

Body dimension (see Fig. 20.2)		Male 95th percentile in	Female 5th percentile in
A	Body height sitting	39·25	31·50
B	Eye height sitting	33.75	27·25
C	Elbow height sitting	11·00	8·25
D	Thigh height sitting	6·75	5·00
E	Abdominal depth sitting	12·50	8·00
F	Buttock–knee length	25·50	20·75
G	Anterior arm reach	37·50	28·50
H	Knee height sitting	23·50	18·00
I	Popliteal height sitting	18·00	14·00
J	Buttock–popliteal length	20·75	16·75
K	Shoulder height sitting	25·25	19·25

The values quoted relate to unclothed men and women since reliable body measurements are impossible with clothed subjects.

Clearly a seat designed to cope with both sets of dimensions would be suitable for 95% of all men and 95% of all women. The dimensions given are for U.S. men (who are generally larger than British men) and for U.S. or British women (who are about the same size) and comparison with other tables of similar information provided by other workers, e.g. Murrell [20(d)], Damon, Stoudt and McFarland [20(e)] suggests that they are slightly high.

It is not possible to use one simple set of body dimensions for any design problem because the body dimensions, reach or power that can be exerted will differ from one body position to another. We need completely different information about a man standing from that for a man seated. Quite a lot can be done at an early design stage, with simple information and common sense, but eventually it will be necessary to turn to more comprehensive information [e.g. 20(e)].

The shape of a man and the movements he can make are not the only human properties that we have to consider. An environment may be hostile even if it fits a man's shape and the designer must ensure that any man in the system is, at least, kept alive and preferably kept comfortable.

One obvious environmental factor to be considered is thermal. If the pilot of an aeroplane were too hot or too cold he would lose efficiency and the success of the plane's mission would be jeopardized. A gold mine, considered as an engineering system, would be useless if the miners were so hot that they were incapable of using the equipment provided. There have been many attempts to define environments that are thermally comfortable and the index most commonly used is that of 'Effective Temperature'. This index considers the dry-bulb temperature of the air, the humidity of the air and its velocity over the man's body and describes combinations of these parameters which are comfortable. The 'Effective Temperature' index is well described in the *ASHRAE Guide* [20(f)] but there is considerable evidence that 'Effective Temperature' is a bad index although its shortcomings are not apparent in conditions near to normal room conditions. Where great departure from normal room conditions is possible, as for example in a supersonic fighter aeroplane cockpit, more sophistication is needed in the calculation of thermal comfort. An acceptable method of determining the thermal comfort of a human environment is to calculate the actual heat exchange at the men's skin, assuming the skin temperature to be 33 °C. This method, described by Billingham and Kerslake [20(g)] has some disadvantages but is much more easily justified than the 'Effective Temperature' index.

Acceleration, vibration and noise are other related environmental factors which have to be considered if a man is part of the designer's system. The acceleration that the crew of a spacecraft can stand (say about 8g) during launch may well be a significant design constraint but even so it will be considerably more than the acceleration to which we would dare to expose an elderly lady passenger of a civil aeroplane. Figures for people have been quoted by MacFarlane and Teichner [20(e)]. The effect of vibration on people is less well understood although everyday experience tells us that vibration is uncomfortable. Again some figures have been quoted by Murrell [20(d)]. Noise has been the subject of much research in recent years and it is well known that constant exposure to loud noise (as might be met in a machine shop or riveting shop, for example) causes a gradual deterioration in hearing. Murrell [20(d)] describes the effects of noise in some detail and also methods of measuring noise but Fig. 20.3 briefly summarizes some of the significant noise levels.

The senses of sight and touch should not be ignored in any man who forms part of a system. Usually the man is there to interpret information that is presented and to act on it. The driver of a car, the pilot of an aeroplane, the power station employee at a console, are all expected to read dials and make adjustments and are essential parts of the transport or power-generating system. It is clearly important that necessary information be presented without ambiguity and that unnecessary information be suppressed. There have been several aeroplane crashes attributed to the pilot's incorrect reading of the altimeter and in some cases it has been shown that the altimeter face has been so designed that the hands on the dial were either difficult to interpret or easy to misread—less a pilot error than a design error.

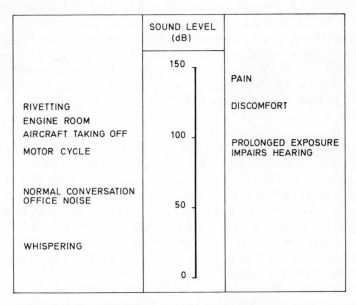

Fig. 20.3 Some approximate noise levels and effects

Operating levers, knobs or control wheels can often seem straightforward to the designer but difficult to an operative in the dark, in the cold, in a cramped position, in a safety belt or in any one of a hundred inevitable situations not considered by the designer. A switch that looks sensible on a drawing may be very difficult to move when the operative is wearing gloves or easy to confuse with another, adjacent similar switch which performs a completely different function.

The designer can avoid unnecessary expense of operation if he considers the operation of the system in detail while the system exists only as drawings. One common way of doing this is to make the designer responsible for producing the handbook of operating instructions. In the aircraft industry the pilot is invariably consulted at all stages in the design of the cockpit and, while mistakes are still made, this does ensure that operating problems are given consideration. With more expensive aeroplanes a mock-up of the cockpit is made so that comfort and ease of operation can be tested before much money is spent on the final article. Often, however, operating difficulties are found too late, as was the case in the following example discovered by a student in the course of project work.

A factory recently installed a newly designed cable-making machine which made cable faster and in longer lengths than earlier machines. Only when the machine had been installed and operated was consideration given to the method of paying the operatives and only then were the methods of operation analysed. It was discovered that while the machine did, in fact, make cable faster than earlier machines (when it was running) no thought had been given by the designer to reloading it with raw material. A reloading system had been improvised by the operatives and subsequently a better procedure was devised after a work study exercise, but much quicker and easier procedures still would have been available had the designer thought of the problem early enough.

With large, expensive systems the ergonomic problems associated with console design and operator comfort are usually considered formally and although specialist ergonomists are frequently employed for this work, the designer himself can cope with most ergonomic problems provided that he makes himself aware of them and uses normal commonsense. Obviously the specialist ergonomist is necessary in certain cases but usually awareness of the problem is all that is required to achieve an acceptable end product. Once again, the best design discipline is simply to devise an operating procedure and go through it, in the mind, in all operating conditions, while the system exists only as drawings.

If mistakes have been made and have not been put right before procedures have become established on the shop floor it may never be possible to correct them. This is partly because the best procedure can often be used only if it has been considered at the design stage and the system is geometrically suited to it, but it is also because it is very difficult indeed to negotiate new payment methods and operating procedures with the operatives, once a slack operating procedure has become established. In a student's study of the method of parcelling and loading scrap in a London factory, a much better and cheaper method of operating was devised but it was not found possible to implement the study because agreements previously made between the firm and its employees would not permit it.

In addition to the ergonomics of manufacture and operation, ease of maintenance must be considered. Maintenance is usually a necessary and expensive part of any system and every motorist knows some maintenance jobs that can only be attempted by long-fingered dwarfs. Once again, the problem is one of prediction. The designer must plan maintenance and devise procedures for maintenance while the system is still being drawn.

Good design management can only occur if ergonomic studies are formally part of the designer's work. This can best be assured by making the designer responsible for publishing the operating and maintenance procedures and by including the supply of the appropriate manuals as part of the contract with the customer. The problem of manufacture is more difficult because, by custom, manufacturing procedures are devised by the planners who may well be out of the designer's control. If this is the case then the only possible safeguard is good, two-way, formal and recognized communication and co-operation between designer and planner.

Exercises and Subjects for discussion

1. Compare the effort involved in different ways of presenting a lecture using for example blackboard and chalk, an overhead projector or closed-circuit television. Could PMTS be used to provide data for a comparison?

2. List some common repair problems on a motor car. How would you expect the tools used by the home owner to differ from those used by the garage? In any of the jobs listed, could better design have made the repair easier and cheaper without any other sacrifice?

3. Consider a car designed as a household's 'second' car. How and where would you accommodate parcels? Sketch a possible car interior.

4. Sketch a lecture room that would be suitable for lectures in mechanical engineering subjects.

5. Compare and criticize watch and clock faces. What other ways of presenting numerical information are there? What should a cricket scoreboard show and how? What information should be supplied to the driver of a family car and how should it be presented?

6. Consider and discuss the housewife's working space. What should a cooking-stove look like? How would you design, and where would you put the control knobs?

7. List some situations in which the environment of an operative must be controlled to make it tolerable or comfortable.

Bibliography

20(a) H. Maynard, *Industrial Engineering Handbook*, McGraw-Hill, New York, 1963.

20(b) F. Neale, *Primary Standard Data*, McGraw-Hill, Maidenhead, England, 1967.

20(c) S. Kirk, E. Edwards and T. Hindle, *Designing the Driver's Workspace*, Design 188, Council for Industrial Design, London, August 1964.

20(d) K. Murrell, *Ergonomics*, Chapman & Hall, London, 1965.

20(e) C. Morgan, *Human Engineering Guide to Equipment Design*, Sponsored by Joint Army–Navy–Air Force Steering Committee, McGraw-Hill, New York, 1963.

20(f) *ASHRAE Guide and Data Book for 1965 and 1966*, American Society of Heating, Refrigeration and Air Conditioning Engineers, New York, 1965.

20(g) J. Billingham and D. Kerslake, *Institute of Aviation Medicine, Scientific Memorandum no. 23*. Farnborough, Hants., 1960.

APPENDIX
VALUE ENGINEERING

Various definitions of Value Analysis or Value Engineering exist. They are all paraphrases, one of the others, and the following definition was published by the British Productivity Council (Notes for a half-day conference in 1969 by J. F. A. Gibson).

'Value Engineering is an organized and systematic effort to provide the required function at the lowest cost (consistent with specified performance and reliability).'

I have added the brackets, which are not in the original definition, because one would expect the 'required function' to include 'specified performance and reliability'. This definition could be a definition of any engineering designer's job. Certainly this book has assumed it to be so but the fact that value engineering is not only fashionable, but has been very useful, suggests that few designers have thought of their work in terms of the above definition.

Value engineering may be applied before production drawings are released to the shops so that an attempt is made to eliminate unnecessary cost before metal is cut, but the more spectacular successes of value engineering result from its application to engineering products that have already been built. Sometimes the term 'Value Engineering' is used to describe cost-saving exercises before building hardware while the term 'Value Analysis' is used to describe cost-saving exercises after some versions of the hardware have already been built. Usually, however, Value Engineering and Value Analysis are terms used indiscriminately, one for another.

Minimizing cost before samples of the product have been built is the task of the designer, as described in this book. There is no doubt that potential savings are greater when only drawings exist than when hardware has already been built. It is, nevertheless, much easier to discover unnecessary cost when hardware has been built.

Value engineers tend to be concerned with the detail of design, rather than the complex system. Their work is not necessarily limited in this way but there is usually a tendency for more immediate savings to be made by attention to detail.

THE VALUE ENGINEER'S PROCEDURE

Selecting the Project

Value engineering requires work, work takes time and time costs money, so that it is necessary for a value engineer to work on projects which will yield a profit. Luckily, it is often found that a few products are responsible for most of the unnecessary costs in a factory so that, early in his career, a value engineer can pick projects which are likely to show impressive results. Circumstances vary from company to company. In one company, it may not be thought to be worthwhile to implement a Value Engineering exercise unless the savings are at least £500; another company might only implement an exercise where the possible savings are at least 20%. In every company, however, there will be a level of saving below which the cost of work exceeds the expected saving. Generally, the Value Engineer will study products where unit costs are high and the volume of sales is large.

Defining the Function

The Value Engineer must decide what the product is for, before he can suggest improvements. This process of defining the function of a product exactly parallels the designer's job of writing a design specification. Usually the designer is concerned with a large, complex system whereas the Value Engineer will look at smaller parts of a system and in this respect, at least, the Value Engineer's job is easier.

The Value Engineer must also decide how much the function is worth. This is a difficult job but can often be done by studying what other people are paying for the same function.

Obtaining Information

Before useful changes can be made, the costs of the existing product must be studied in detail. The value Engineer must therefore obtain and study:
the product itself, if it exists;
competing products, if they exist;
a set of manufacturing drawings;
raw material specifications and costs;
all relevant process schedules, with manufacturing times;
labour costs; inspection and scrap histories.

The designer also needs this information but at the stage at which a designer creates an idea, information of this sort can only be prediction.

Create Cheaper Solutions to the Problem

By some method such as brainstorming, it is necessary to create possible solutions to the defined problem.

This again exactly parallels the designer's creative function although the Value Engineer usually enters the field later than the designer and can usually criticize ideas already put forward by others.

Analysing

Each of the ideas proposed must be analysed and either accepted or rejected. This again is a function exactly parallel to that of the designer. Ideas must be checked for feasibility and, if feasible, compared on a cost basis. The best is selected.

Implementing

When an idea has been shown to be good, it should be implemented. This is the most obvious but usually the most difficult of the Value Engineer's jobs. The Value Engineer must be sufficiently senior in the organization to ensure that good ideas are implemented.

Reporting

When a Value Engineering exercise has been carried out, it should be followed up and costed to see if the predicted savings were in fact, achieved. An honest report should be issued to say what was done, why it was done and whether it was successful.

THE VALUE ENGINEER'S TEAM

The Value Engineer obtains information and makes suggestions but he is mainly the co-ordinator of the work of others, whose expertise he needs.

Usually the Value Engineer holds regular meetings which he chairs. The team will consist of:

the Value Engineer—in the chair;
the Chief Designer;
the Chief Production Engineer;
the Chief Buyer;
the Chief Estimator;
and any other senior man with knowledge to contribute.

Titles change from one firm to another but the areas of knowledge are clear and seniority must be such that what a Value Engineering Committee decides will be implemented.

USEFUL QUESTIONS FOR THE VALUE ENGINEER TO ASK

What is it for?
Do we need it?
Is someone buying the same function for less money?
Is any part of it redundant?
Can it be made of cheaper material?
Can it be made by a cheaper process?
Can standard parts be used?
Can we use cheaper or better tools?
Can we buy the same materials cheaper?
Can we save some paperwork?
Can we widen tolerances?
Can we buy in better quantities?
Can inspection be simplified?

ROADBLOCKS

Many people have set ideas (usually unjustified) which prevent argument.
'The men (or the boss) won't wear it'. Try asking them or, better still, offer them some of the profit. They might enjoy co-operating.
'We've always done it this way'. Perhaps it's time for a change.
'There are technical reasons for not changing'. What are they?
'We have already looked at this product and no improvement can be made'.
If true, history is full of cases where two consecutive studies of the same problem both show improvement.
And so on.

What is required is an open mind, a knowledge of the facts and creative thinking.

FALL-OUT

Value Engineering is intended to save money but there are often secondary benefits.
Reliability, Maintainability and Weight are improved in nearly half the cases where Value Engineering is applied.
Logistics, Performance and Customer Appeal are often improved.
Factors connected with production such as Ease of Production, Lead Time and Parts Availability are usually improved.

SOME SIMPLE EXAMPLES

Fibreglass Tape

A company making aluminium extrusions used $1\frac{1}{2}$ gross rolls of 1 in fibreglass tape per month in packing its products.

The original cost of the tape was 86p per roll.

Because the company bought so many rolls of tape the buyer thought it worthwhile to negotiate a special price. He found that he could obtain the rolls at 65p each.

Tests showed that $\frac{3}{4}$ in tape performed well and then, that $\frac{1}{2}$ in tape performed well so that, eventually, $\frac{1}{2}$ in tape at $32\frac{1}{2}$p per roll was used for packing.

The cost of the exercise was small.

The cost of implementing was negligible.

The saving was

$$12 \times \frac{144 \times 3}{2} \times 53\tfrac{1}{2}\text{p} = \text{£}1300 \text{ per year.}$$

Rotating Switch

A company made a successful switch, proved on prototype aircraft. Receipt of an order for a large quantity of the switches suggested that a Value Engineering study would be justified. This study resulted in the saving of £12·50 per switch. It is worth noting however that, of that £12·50:

£3 per switch was saved by bulk buying and

£5 per switch was saved by modifying the final inspection procedure;

thus £8 per switch was saved with no design changes and at negligible cost of implementation.

The total saving on the whole order was £3600 and this was all profit.

Butterfly Valve

A company made a butterfly valve operated by a bought-out, electric actuator. When the early valves had been made it was discovered that the cost of manufacture exceeded the originally agreed price so that a Value Engineering exercise was suggested.

The following resulted:

£12·75 was saved by buying a similar actuator from another supplier.

£12·50 was saved by using a cheaper limit switch.

The original limit switch was the only one advertised to meet the environment but the company's reliability records showed that a cheap switch had survived the environment and could be used.

The original design called for a non-standard keyway and key. This was replaced by a standard keyway.

An X-ray of every casting had originally been requested by the drawings. This was found to be unnecessary.

Minor changes to details and the elimination of some tests reduced costs.

Savings were £40 per valve and resulted in profit.

GENERAL COMMENTS

Value Engineering is fashionable and it works. Often, however, the use of Value Engineering merely reflects a realization, on the part of the management, of the importance of costs.

After a few years, almost any company will settle into bad habits and so almost any keen Value Engineer will make impressive savings early in his career. Value Engineering effort and results do tend to peter out after a couple of years. This is partly because the obvious savings have been made and partly because enthusiasm wanes. The manager has to decide whether to apply Value Engineering at discrete intervals or whether to back a constant effort.

Value Engineering has the same objectives and many of the same techniques as Engineering Design. The perfect designer would make the Value Engineer redundant.

Index